SCRAP METAL RECOVERY
An experience of intermediate

technology in Papua New Guinea

Colin Relf

IT Publications 1986

Intermediate Technology Development Group (ITDG)
Myson House
Railway Terrace
Rugby CV21 3HT, UK.

ISBN 0 946688 57 5

Printed by Russell Press, Nottingham, England.

Colin Relf is a freelance development consultant who has been involved in a wide range of work including analytical studies of development policy and programmes as well as the planning, appraisal and evaluation of specific development projects. He has worked extensively in Africa and Asia with international agencies, developing country governments and non-government organizations.

ITDG would like to thank the Scottish Catholic International Aid Fund (SCIAF) for their financial assistance towards this publication.

Photos: © ITDG.

Drawings supplied by the South Pacific Appropriate Technology Foundation (SPATF) Papua New Guinea.

Cover photo: Molten aluminium is drawn out of the furnace into the waiting moulds.

.

PREFACE

When the peoples of Africa, Europe and Asia first began to develop foundry skills, their craftsmen were believed to have magical powers. And still today there is a fascination, even a thrill, in watching silver rivulets of molten metal flowing from the glowing heart of a furnace and hardening in a mould to produce some useful product. For us in the South Pacific Appropriate Technology Foundation (SPATF), the thrill is heightened for a number of reasons. In the Hanuacast Foundry our raw material is scrap metal. Through using a technology which is appropriate to the conditions in Papua New Guinea, we are turning this scrap into new forms which can be exported and into new products which are needed in our country and which otherwise would have to be imported. In the process, we have also created a new chain of employment opportunities — not only to the 'magicians' in the foundry itself — but also to the much larger numbers of people who collect, sort and deliver to us our supplies of scrap.

For all these reasons, it makes me very happy to see, as marked by the publication of this book, the story being told of this 'magic' conjured up by the wonderfully helpful and productive relationship which has grown between ourselves and ITDG over the past five years or so.

It was in 1981 that ITDG provided SPATF with the 'grandfather' of the foundry in the form of Israeli expert, Willie Feinberg, who built our first experimental furnace. And believe it or not (we could not!), one of the main ingredients in making that furnace was cow dung.

With funding support from the USAID-backed Foundation of the Peoples of the South Pacific and with

(v)

technical and managerial input from USO volunteers Steve Layton and Peter Thomas, the foundry then grew from nothing until, by 1984, it was turning over about K100,000 a year.

Early in 1985, the 'Grandfather' was sent again from Israel at the behest of and with funding secured by ITDG, to work alongside Wilson Amos, Peter Thomas, Pepe Pukari and other foundry workers to build a large version of the furnace more capable of handling the high volume of scrap and producing more castings.

This was achieved with generous funding from such sources as the New Zealand High Commission and the Scottish Catholic International Aid Fund (SCIAF). The latter agency having also provided funding to assist with the publication of this book so that others too may learn from our experiences, success and mistakes.

Integrated into our Small Industries Research & Development Centre here in Port Moresby, and with the added impetus of Mel Johnson, an Australian volunteer who has come to train the workers in pattern-making, the foundry is now producing a rapidly increasing range and number of castings which are essential for import substitution and the development of local industries in our country. For example, just in the past month we have taken orders from the World Health Organization to cast 300 aluminium heads for the Blair hand water pump, we have completed an order for artificial limbs, and we have been asked by one of the largest local engineering firms to start manufacturing pulleys for them since 'the quality of your castings is, if anything, better than what we get from Australia and they cost less'!

But overall, the foundry has been a tremendous success story and indirectly provides employment for dozens of people who might otherwise have no source of income. It is even becoming highly profitable now that past problems of scrap thefts have been solved.

For all this, we are very grateful to our partners in ITDG and to Colin Relf for having produced such a masterful and detailed study of what we are trying to achieve and how we are trying to attain it. I would only add that this book tells the story of Hanuacast up to the beginning of 1985 and that, since then, the foundry has continued to learn and to devlop. For those who would like to know the latest of what we have learnt and how we have developed, there is only one way — come and see for yourselves.

Andrew M. Kauleni

Executive Director
South Pacific Appropriate Technology Foundation

CONTENTS

(Approximate exchange rate: £1 = K1.25)

INTRODUCTION

To read almost any selection of economic reports on less developed countries is to gather the impression that there are myriad interesting and productive possibilities for development if only . . . if only farmers had access to the right inputs; if only experienced small-scale entrepreneurs could be found; if only this technology or that modest amount of capital could be made available; and so on. In practice such conditions are met quite rarely. Even when they are or can be met the original analysis of prospects all too frequently is shown to have been incomplete or to have paid inadequate attention to the equally numerous pitfalls before success can be achieved.

In Papua New Guinea, however, there is a modest success story. It is a small non-ferrous metal foundry using simple, cheap technology which has been in successful operation since late 1981. In its second full year of operation its throughput of scrap metal — all of which was sold — was *at least* 130 per cent greater than during its first year. In its third year it continued to grow. This report is an attempt to tell the story of this, the Hanuacast foundry, to describe its current operations and to suggest what may be ahead for its further development.

The Hanuacast story also starts with an "if only . . ." It is perhaps not commonly appreciated that in a developed economy such as the United Kingdom the proportion of re-cycled scrap in new copper products is 40 per cent; in the case of lead products scrap represents about 60 per cent by weight. Papua New Guinea has a dualistic economy with a strong modern, urban-based sector using modern technologies and materials. These generate scrap. At the same time Papua New Guinea's own history and culture has never involved any experience in the extraction, smelting and refining of metals. Ergo, there should be plentiful supplies of non-ferrous scrap and *if only* an appropriate technology could be devised and made available there should be a good possibility of collecting, selling and perhaps using this scrap.

On the strength of this basic analysis, in 1981 the

Intermediate Technology Development Group responded to a request from Papua New Guinea to provide a technical adviser in foundry work and non-ferrous metal casting techniques. This consultancy appeared to be successful at the time yet after a while little news filtered back to ITDG other than a confirmation that metal smelting work was still going on.

In February 1985 ITDG felt that it was time to gather some more complete information. This report is the result of that decision and responds to the following terms of reference:

(i) to review implementation of the (Hanuacast) project from its original conception to date, and describe the current operation of the project, its organization, activities and techniques;

(ii) to explain the changes which have taken place in the project's objectives and in the nature of its activities, products and organization, identifying reasons for the changes and how they were brought about;

(iii) to assess the effects of the project on employment and incomes, directly and indirectly, and the distribution of the benefits among socio-economic groups;

(iv) to appraise the likely social and economic impact of the proposed expansion of the furnace upon those currently involved, directly and indirectly, in the project's activities.

The visit to Port Moresby took place between February 25th and March 10, 1985. During that time information and opinions were gathered from a wide range of people, including government officers; the staff of the foundry and its parent organization the South Pacific Appropriate Technology Foundation; academics; and a sample of those involved in collecting supplies of scrap metal for the foundry. The willingness of people to give so freely of their time and to be so open with information was partly a reflection of the natural generosity of spirit that is a feature of Papua New Guinea. But it was also partly that everyone who is involved with or has heard of the Hanuacast foundry finds it a fascinating and exciting project. All are eager for its success

(x)

to continue and for its story to be told.

Among less developed countries Papua New Guinea has some unique features and among industrial enterprises within the country Hanuacast is itself unusual. This report starts therefore with an extensive background section on PNG's experience with industrial development so far and the current policy framework for new industrial growth. In this way, as the Hanuacast story unfolds, it is hoped that the fact of its success and its potential role as a paradigm of small-scale labour-intensive manufacturing can be appreciated in context.

Hanuacast is still growing, still learning, still developing. During this investigation itself work was under way to transfer the foundry premises to a new location with more space and better prospects for improved layout. And there are many ways in which its learning process and market opportunities might lead it. At the same time the foundry has experienced and still is experiencing a number of deficiencies and problems.

Various types of scrap metal await further treatment at Hanuacast.

I THE SETTING

Background

Papua New Guinea is a country of sharp contrasts, not least because of the broad social and economic transformations it is still going through some 10 years after gaining independence. Though it is classed by the United Nations as a less developed country there are some characteristics, especially of the rural economy, which fall far below the level of development that even this classification suggests. In a number of other respects, however, such as living standards in the major urban centres and in the applications of modern technology in some sectors, it is considerably in advance of many other less developed countries.

As might be expected, more than three-quarters of the population live in scattered, often extremely remote subsistence communities. In some areas, subsistence means little more than hunting, fishing and gathering. Even in some relatively fertile areas agriculture has not developed beyond a rudimentary form of shifting cultivation with family "gardens" of many different interplanted crops. Papua New Guinea must also be one of the few countries — if not the only one — whose capital is not linked by road to any other urban centre of importance on the same land mass. The country's secondary school enrolment ratio is only about 12 per cent. These are a few examples of how far PNG has to go in physical, social and economic development.

In contrast, the remote rural areas are served by an extensive network of about 450 airstrips. PNG's rapidly spreading telecommunications system has the latest microwave technology. It has a highly developed and capital intensive mining industry contributing about one twelfth of value-added to GDP (but accounting for less than one-fortieth of total employment). And in contrast to the very low levels of per capita income among subsistence communities, minimum wages in PNG's manufacturing sector in 1978 were almost 12 times greater than in Indonesia, about 8 times higher than in Sri Lanka and 4 times more than the urban minimum wage in Thailand.

1

The Pattern of Development in Manufacturing

It is only in the last twenty years or so that there has been any significant industrial development at all in Papua New Guinea and the manufacturing sector in particular is at a very early stage. Not surprisingly, the lack of any traditional artisanal manufacturing production means that the structure, types of activity, geographical spread and even the technologies used in PNG's new manufacturing industries are quite different from those in many other less developed countries. Table 1 (page 7) outlines the basic structure of the manufacturing sector in 1981, the most recent year for which national data are available.

Most noticeable, perhaps, is how small the manufacturing sector still is, with less than 500 factories and only about 21,000 people employed. Nevertheless, it is a reflection of the country's economic dualism and the small size of the "modern" sector as a whole that even this modest level of employment represents about 10 per cent of all formal employment. Indeed, manufacturing is already the third largest employer (after Government and public authorities, and primary industry). It also constitutes about 10 per cent of total value added in GDP.

With an average of 44 persons employed per factory, manufacturing units in PNG are relatively large. In most less developed countries the large number of small-scale enterprises employing less than 5 people tend to reduce considerable average employment per unit. In Papua New Guinea, however, factories employing 1-4 people in 1979 accounted for less than 14 per cent of all factories and for only about one per cent of all manufacturing employment.

A report from the National Statistical Office shows an average value added per person employed of nearly K10,000 per annum for the sector as a whole. In the case of the chemicals sub-sector and the food, beverages and tobacco group, high levels of value added might be expected. But even such industries as ISIC groups 32 (textiles, garments etc) and 33 (wood, wood products, furniture, etc) which in many countries are typically small-scale and labour-intensive

2

show quite high average employment per factory in PNG. And although the value added in production in these subsectors are the two lowest, at K6,200 and K6,400 per annum respectively, they are still high in comparison with other less well developed countries. Certainly, in relation to the urban minimum wage in 1981 (equivalent to K1428.5 for a 50-week year) these levels of value added do not signal the use of labour-intensive production techniques.

Analysis of growth trends in manufacturing is difficult partly because there have been changes and inconsistencies in data collection methods and partly because a backward look over even a short period shows a rapidly diminishing base, hence misleadingly high growth rates. What can be said, however, is that over the period 1970-81 GDP grew at an average of about 2.5 per cent per annum while the contribution from the manufacturing sector was considerably higher at about 7 per cent per annum.

Manufacturing has also shown strong growth in employment — at least in relative terms. Between 1976 and 1980 total formal sector employment grew at just over 5 per cent per annum: in manufacturing and utilities the corresponding growth was more than 8 per cent per annum. In absolute terms, however, this growth represented only about 1,400 jobs per year on average. In contrast, there were about 40,000 new entrants to the labour force in 1980.

The lack of a traditional base of small-scale community-level manufacturing and the characterization of PNG's industrial development as recent, urban-based and high-technology-biased are borne out by the results of an industrial census in 1979. This showed that the country's two largest urban centres, Port Moresby and Lae, accounted for 43 per cent of all manufacturing enterprises and 50 per cent of all manufacturing employment.

At present it is also clear that the manufacturing sector is principally serving domestic demand. In 1982 only about 10 per cent by value of manufactured output was exported and contributed less than 4 per cent of the value of total exports. The two main exported items were copra oil, and plywood and other wood products which together accounted for over

90 per cent of all manufactured exports. Moreover, what manufacturing there is in the country is heavily dependent on imported inputs, including imported commodities required for the production of finished goods. In 1972-73 the value of imported inputs in manufacturing accounted for 56 per cent of all intermediate inputs while in 1976-77 this figure had risen to 71 per cent.

In sum, Papua New Guinea's manufacturing industry is at present very much at an infant stage. The number of factories is small and the range of products they make is also limited. And because there was historically no manufacturing base until recent times, there has been little experience in the country of the adaptive development of technology. Instead, manufacturing enterprises have tended to be "planted" in PNG, making products and using technologies that were developed in quite different economic environments.

The Informal Sector and Urban Poverty

If manufacturing is at a very early stage of development in Papua New Guinea the informal sector is hardly evident at all. In 1980, although Port Moresby's total population was about 124,000 there were only two places in the city offering shoe repair services. Unlike the urban centres of other less developed countries, PNG's towns have very limited numbers of back-yard vehicle repair and other small workshops. In Port Moresby there are only a few tailor shops, they are mainly in modern shopping arcades, and their sewing machines are electric, not hand-operated. There are no special parts of the city where people can be found making, say, cheap sandals, cooking utensils or other household items: there are no street-corner shoe shine services; no street vendors; and instead of sweets, flowers, matches, chewing gum and so on, the only thing that children are likely to offer for sale to motorists at traffic lights is a newspaper.

There are various reasons for this very low level of informal sector activity. Most fundamentally, there is little going on in the informal sector now because there was little going on before. Historically there has been little specialization of skills in Papua New Guinea's villages. All the skills needed for subsistence were passed on within the family unit. And when urban development began relatively recently these skills were not really needed in the towns. One of the few exceptions was wood carving, a traditional skill whose products did find a "modern" market. Indeed, wood carving remains the largest category of (mainly part-time) informal sector employment.

When the colonial administration came, virtually all the material goods required and the skilled manpower to maintain and repair them were literally shipped or flown in and the "modern" sector was superimposed wholesale. The towns were established principally as administrative centres and trading posts. With little diversification in the economy there were few opportunities for acquiring skills required for small-scale manufacturing or for self-employment in an informal fringe.

Another factor which has surely suppressed the growth of informal sector employment is the relatively high cost of living in the urban centres of PNG. The cost of food and clothing is little different from what it is in Australia; housing is cheaper; but furniture, electrical goods and other household items are considerably more expensive. Of course, the relatively capital-intensive production techniques in factories and the high average value added to materials outlined earlier have enabled the manufacturing sector to afford to pay quite high wages, the current legal minimum standing at K45.2 per week. But these high wages and the high cost of living in towns have left little scope for people to be involved in labour-intensive, low income fringe activities and to survive.

Government regulations have also actively constrained the development of an informal sector. Since the early 1970s a number of restrictive regulations have been in force which are very similar to those applied in Australia. Planning permission is required for all new buildings and for the uses to which they are put. There are comprehensive factory regulations governing safety aspects, the provision of running water, toilet facilities and the like. In 1973, the Chief Minister in the pre-independence government agreed in principle to relax some of those regulations, saying:

"We will expand small-scale commercial activity. We will cut down the restrictions that prevent people from earning money by selling goods, supplying services or engaging in small-scale businesses. Our current licensing regulations are borrowed from a Western-style economy. They may have no meaning in an independent Papua New Guinea".

So far, however, there are no signs of any relaxation in these inherited regulations. On the contrary, the particular reason for the absence of street vendors in Port Moresby is a by-law introduced by the city authorities in December 1983 explicitly prohibiting the sale of goods in the streets.

Nevertheless there is poverty in the towns and there are people living on the fringe of the high wage, highly regulated urban economy. The 1980 census showed that there were 7,740 urban households without a wage earner in the country as a whole, representing 13.6 per cent of all urban households. Subsequent research has shed some light on how these households derive their incomes, and how much worse off they are than those with a regular wage income.

Table 1 shows the results of a 1983 study of the Nine Mile and Gordons Ridge areas, two low income settlements in Port Moresby. In both settlements the majority of households do have at least one wage-earner. At Nine Mile the main source of wage employment is a quarry while Gordons Ridge is more centrally located between the airport and the main government centre at Boroko and has a more diverse pattern of employment. In both settlements, however, transfers are the second most important source of income. This is consistent with the frequently quoted "wantok" system in which mutual support is provided not just within an extended family but within a clan or sub-clan group which can often be defined in terms of a shared language. Food marketing, retail trading and "other" sources are those which can be associated with the informal sector. Yet for the two settlements together they are the main source of income for only about 11 per cent of all households. It is also interesting that food marketing is an important main source of income. It is a particular feature of low-income urban settlements in PNG that the settlers (or squatters as they are frequently called) often cultivate extensive food gardens both for home consumption and for sale. This, along with wood carving, is another traditional occupation which can still be pursued in an urban environment provided there is enough land.

Table 1
Main Sources of Cash Income from Two Low-Income Settlements in Port Moresby, 1983

	Nine Mile	Gordons Ridge	Total
	(Per cent of households)		
Wages	69.0	78.7	75.6
Transport business	0.0	0.0	0.0
Retail trading	3.0	0.0	1.0
Food marketing	5.0	2.9	3.6
Transfers received	20.0	9.7	13.0
Other	3.0	8.7	6.8
Total	100.0	100.0	100.0
Number of households	100	207	307

The types of income-earning activities included in the "other" category in Table 1 are quite limited. They include collecting and selling firewood; collecting, sorting and selling bottles; fishing in the coastal settlements; and in more recent times, of course, collecting and selling scrap metal. Only in a few cases do these informal activities involve special skills. One example is a group of men in the Gordons Ridge settlement making cane chairs — a skill they had learned from missionaries before leaving their home village.

In some settlements there appears to be a considerable gap in incomes between households with and without a wage-earner. At Nine Mile settlement the average income of a sample of households with a wage-earner was just over K126

7

per week in 1983. For a similar sample of households without a wage-earner the average weekly income was under K29. At Gordons Ridge the average incomes were about K58 and K25 respectively. These averages do not, of course, take into account that there are many single men living in squatter settlements who are either not married or have left their immediate families in the home village. But additional data on income per head in households with and without wage-earners still show significant differences.

This brief review has attempted to show that the urban informal sector is limited in scale and scope partly because of a lack of skills and entrepreneurial capacity among the worse-off people and partly because extensive government regulations have restricted the opportunities. Meanwhile there is a relatively small but nonetheless significant group of people, most of whom have resettled in the towns, who have not been able to find formal employment, who rely to a great extent on income transfers from kinfolk, but whose incomes are frequently way below those who do have jobs.

2 THE HANUACAST FOUNDRY

The Foundry in Context

Hanuacast, based in one of Port Moresby's "urban villages" is a relatively new manufacturing venture. How, then, does it fit against the background of industrial development so far? To what extend has it been affected by the constraints commonly identified for domestic manufacturing operations? And what have been the effects on Hanuacast of the government's own efforts to promote new industrial development? The general answer is that Hanuacast does not fit well in the mould of analysis in the previous section. For example:

In Hanuacast's production process:
  (i) with a foundry staff of only 13 people it is considerably smaller than the national average for manufacturing enterprises;
 (ii) it uses a low level of technology — certainly one which has not been imported from a more developed economy;
(iii) nevertheless, it achieves a rate of value added to materials per foundry employee per annum of between K15,000 to K17,000, which is considerably higher than the national average of about K10,000 in all manufacturing;
 (iv) although all the raw materials it uses originated overseas, they cannot be regarded as imported in the sense that they do not add to the nation's net import bill;
  (v) unlike most PNG manufacturing, the bulk of the foundry's output is exported;
 (vi) Hanuacast has not involved any foreign investment, nor has it been 'planted' wholesale but has had to adapt and evolve within the domestic environment;
(vii) unlike most manufacturing operations the foundry not only has strong links with the informal sector but relies fundamentally on it.

In relation to the common constraints affecting the

viability and growth of manufacturing:
 (i) Hanuacast is not affected by the country's limited infrastructure and its catchment area for raw materials could not expand greatly beyond the urban limits of Port Moresby even if the road network were more extensive;
 (ii) it is not affected in its present type of operation by the fragmentation and limited purchasing power of the domestic market;
 (iii) because it uses a simple technology and makes simple products it is not seriously affected by the limited skills available in the local labour market.

And in relation to government policies and programmes and the institutional framework for supporting industrial development:
 (i) Hanuacast has not been affected by any of the fiscal incentives (because its status as part of a charitable organization has made it exempt from taxation).
 (ii) it has never applied for nor received any direct assistance through the National Investors Scheme or the Credit Guarantee Scheme.

Origins of the Hanuacast Project

The origins of the Hanuacast foundry lie not in Port Moresby, nor even primarily in ideas about scrap recovery as an operation in itself. Instead, the story begins in Lae, PNG's second largest urban centre and ideas about metal smelting and casting had more to do with manufacturing products for sale in domestic markets and with setting up small-scale industries in the rural areas. Since the present day operations of Hanuacast owe much to the lessons learned in Lae in 1980 and 1981 it is worthwhile tracing the experience in this first venture in metal foundry work.

The very first metal foundry in PNG — the Tru-Cast foundry in Lae — was set up by an Australian entrepreneur. This project was in fact quite typical of the type of manufacturing development characterized in the previous section of this report. It was foreign-owned and foreign-financed; it used an imported technology involving a furnace capable of melting cast iron using expensive electricity as power source; and its aim was to produce a range of cast metal goods for sale in domestic markets. Eventually, it failed.

Meanwhile, in early 1980, local interest in the possibilities of metal foundry work had been stimulated by the Tru-Cast operation. The Appropriate Technology Development Unit in Lae began its own programme of research and development in foundry practice. (ATDU had been created in 1977 as part of the University of Technology in Lae. Subsequently, in 1981, the Unit became an Institute — ATDI — with more autonomy but retaining the same objectives of research, development and promotional work in appropriate technology.)

The Lae foundry project as it came to be known was geared to the manufacture of high quality products, with the initial focus on cast components for micro-hydro units. Using the Lae complex as a production-cum-training operation it was then hoped to establish small foundries in different parts of the country. These aims were fully consistent with the principles of making PNG more self-reliant in manufactured goods, of making products to improve living standards in the isolated rural areas (e.g. village electricity supplies from the proposed micro-hydro units), and of extending the industrial development process itself to the rural areas.

As so easily happens, however, the principles and the philosophies that surrounded the Lae foundry project far outstripped the practical possibilities. The Morobe Provincial Government had approved some modest financial assistance and it was expected not only that orders could be obtained for new products but that they could be profitably fulfilled. Trainees were expected to be able to earn sufficient income from jobbing work during their apprenticeships to set up their village-level foundries when the time came. But a number of crucial aspects had not been fully assessed. Market requirements and prospects had not been thoroughly researched, no analysis of production costs of specific products in relation to the price and quality of imported

11

alternatives, and ATDU itself did not really have sufficient knowledge of production processes and techniques to get the Lae foundry on its feet.

By mid-1980 ATDU realised that outside assistance would be necessary and accepted an offer of technical advice from the Intermediate Technology Development Group. At about the same time ATDU also telephoned a scrap metal processing company, Non-Ferral Pty in Australia, to see if they, too, could give advice on smelting non-ferrous metals. As it happens, these two outside contacts were of crucial importance not so much for the Lae operation but for the subsequent development of the Hanuacast foundry in Port Moresby.

Constructing the aluminium furnace.

The ITDG contact brought a consultant, Mr. Willie Feinberg, to Lae. The terms of reference for this consultancy reflect what were still — at least in retrospect — the rather ambitious expectations surrounding the Lae project. They called for advice in lost wax casting processes for decorative products, micro-hydro components and other engineering castings; the development of formulae for rammable and castable refractory mixes; the manufacture of crucibles; and advice on subsidiary manufacturing operations and processes. The terms of reference did include an item referring to the design of a low-cost furnace for scrap metal reclamation — but the consultancy was only for one month and the priority was clearly on the first, production-oriented set of issues.

The consultant's report confirms that the main focus was on casting materials and techniques and on identifying the types of products that could be manufactured. The report remarks almost in passing that "during the visit additional information was offered in setting up scrap collection and running small-scale industries". In other words, the reclamation of scrap metal was seen at this stage by all concerned as a means to an end — as a source of raw material for the manufacture of finished products.

Neither Tru-Cast nor the ATDU Lae foundry project lived up to expectations. When Tru-Cast failed the Morobe Provincial Government stepped in and bought the operation. But, again in retrospect, it seems that the Province had no clear idea how to revive the foundry but purchased it because it was there and would otherwise have been dismantled or simply abandoned. The Lae foundry project also had its disappointments. In November 1980 — about six weeks before the ITDG consultant's visit — a progress report referred confidently to an order having been received for 5,000 grates from the Department of Minerals and Energy (to be distributed in various parts of the country as part of a programme to introduce more modern cooking techniques). For a new venture a large order of this type would have been great help both financially and in terms of perfecting production techniques, work processing, plant layout and the like. It was also expected that aluminium handles could be cast for bush knives, of which 400,000 were reckoned to be sold each year. But less than three months later, after the ITDG consultant had left, the order for 5,000 grates had failed to materialize and the report concluded that "the first attempt to go into commercial production has been unsuccessful".

It was still expected in Lae that other small foundries could be established — and it had always been planned that one of them would be in or near Port Moresby. It was at this stage, however, that there were two developments which together determined the way in which the Hanuacast foundry would be established.

First, the South Pacific Appropriate Technology Foundation (SPATF) itself became more interested in small industry development. SPATF had all along been closely involved in the Lae work, indeed ATDU was originally a project initiated by SPATF as an attempt to bring the resources of the University of Technology to bear on appropriate technology issues. But SPATF had forged a role for itself in providing information, promoting appropriate technology through demonstration workshops and the like. It was only at the turn of the decade that SPATF decided to become more directly involved in the application of appropriate technology through supporting and fostering small industries which would actually use these technologies. SPATF's new Small Industries Programme therefore had room in it for a small-scale foundry.

Secondly, the earlier contact between ATDU and Non-Ferral in Sydney gave some results when a company representative visited PNG in 1981. Non-Ferral say that they found the staff of the Lae project to be 'disheartened', but there was greater optimism in Port Moresby where SPATF had already decided that a foundry should be included in its Small Industries Programme.

Of course, Non-Ferral's commercial self-interest was in increasing supplies of high-grade non-ferrous scrap. Ever since the Second World War the Pacific basin had been a source of scrap. Initially, much of the supply had been from salvaging abandoned war material though as time went on new construction projects, industrial development and old motor vehicles had also generated a variety of scrap. But according to Non-Ferral, supplies reaching them had been sporadic, usually sent by expatriates on an "as and when" basis, and with no sorting of different types of scrap according to value. From time to time Non-Ferral staff had toured the region, including Indonesia. New Caledonia, Tonga and Fiji, trying to improve the quality and flow of scrap. But its main overseas source had remained the more developed neighbouring economy of New Zealand.

During the 1981 visit to PNG, however, Non-Ferral's message that scrap had value found for the first time a receptive ear at SPATF headquarters. Accordingly it was decided that the new foundry would start modestly, with scrap reclamation as the first objective. The role of the foundry would be to melt down suitable scrap, aim for consistency in the melted mix and produce ingots which would have a higher value-weight ratio and be easier to ship than raw scrap.

The Pattern of Hanuacast's Development

The Hanuacast foundry was set up in late 1981. SPATF made premises available at its Hanuatek complex, a series of workshops with living accommodation in Hanuabada, one of Port Moresby's urban villages. Hanuatek had been built originally by a neighbouring vocational training school. It had been intended as a small enclave where selected students, after completing their course, could spend a period applying their skills under supervision as "fledgling" entrepreneurs before moving on to start businesses of their own elsewhere. SPATF had taken over the complex in the hope that it, too, could launch artisans and entrepreneurs in businesses based on appropriate technology.

It had been planned that staff from ATDI (as it had by then become) in Lae would be responsible for setting up the Hanuacast yard and that two trainees from the Lae foundry would remain in Port Moresby to run the new foundry. In the event, however, the ITDG consultant Willie Feinberg was able to make a second visit and it was he who made the foundry's first two furnaces.

During the original consultancy to Lae a vertical cylinder furnace had been designed using a buried 200 litre steel drum fired by bottled gas. In Port Moresby, however, it

15

Collecting clay for use in building the furnace.

was felt that gas might be too expensive, particularly because it was envisaged that scrap metal would be melted down but still sold as scrap. Waste engine oil seemed to be a viable alternative. Indeed, as far as Hanuacast was concerned the use of this fuel represents a chance — but fortuitous — example of the restrictive government regulations actually being of benefit to a new small-scale venture. In Port Moresby all garages and public agency vehicle servicing depots are forbidden to dump waste oil, yet there is no public collection and disposal services. There are thus plentiful supplies of waste sump and gear oil and Hanuacast was able to procure some of this oil free of charge except for the cost of arranging transport. Two furnaces were installed at Hanuacast with waste oil as fuel. One — a vertical cylinder type similar to the Lae furnace made earlier — has been described in an existing ITDG publication. The second was a reverbatory arch furnace specifically designed to melt aluminium alloys away from steel and other components. A cutaway drawing of this furnace is shown in Fig.1, page 28.

Organizing the collection of scrap presented few difficulties. The two main rubbish dumps for Port Moresby, at Baruni and Six-Mile were easily accessible and there were already settlements of migrant "squatters" living near these dumps, similar to the groups at Gordons Ridge and Nine-Mile. All that was necessary was to spread the word that Hanuacast wanted regular supplies of scrap metal and the prices it was willing to pay. What was more difficult at first was to explain what types of scrap metal would be of value. This took time, not least because the foundry staff themselves were not experienced in recognizing different types of metal in relation to the prices Non-Ferral would pay. During this learning process it is quite likely that Hanuacast would have paid for some scrap which it could not use or paid too much for composite pieces of scrap which needed more work at the foundry before it could be melted down and shipped.

On the receiving end, Non-Ferral say that at first they were not very confident about what, if anything, would come from the encouragement they had given to SPATF in Port Moresby. And in the early days containers were said to appear "out of the blue" and at intervals of up to two or three months. Packing of the freight containers was fairly poor, some of the scrap was quite low grade, and the aluminium ingots were found to contain high proportions of contaminants, particularly zinc. (Non-Ferral are able to make rapid analyses of ingot composition using a spectrograph and the price offered is based in the suitability of the alloy for further refining.)

Perhaps the most graphic indication of Hanuacast's development is provided by the record of its shipments to Non-Ferral. The first recorded shipment in Hanuacast files was covered by a bill of lading dated 5th February 1982 (though it is believed there were some shipments prior to this date). From then to 11 February 1985 there have been 45 shipments — 13 in 1982/13; 17 in 1983/84; and 15 in 1984/85.

Table 2
Summary data on scrap shipments by year

	1982/3	1983/84	1984/85a
Number of Shipments	13	17	12
Total Net Weight (kgs)	73464	167074	142910
Average Weight/Shipment (kgs)	5651	9828	11909
Total Net Value (kina)	37391	110679	91780
Value per Shipment (kina)	2876	6511	7648
Average Value per kg.	0.51	0.66	0.64
Total Freight Costs (kina)	7196	17909	12876
Freight/Value Ratio (%)	19.2	16.2	14.0

a. Data on the remaining 3 shipments from December 1984 to February 1985 not available.

There was a dramatic increase — of nearly 130 per cent — in the weight of scrap shipped in the second year. In the third year (if the average weight of the first 12 consignments was maintained for all 15 consignments) there would be a further increase of about 7 per cent in total weight shipped. There has also been a steady increase in average weight per shipment, reflecting better planning by Hanuacast and increased efficiency in packing the standard sea freight containers.

The value of scrap shipped has also increased. This, though, is the result of four factors two of which have been beyond Hanuacast's control. First, there have been movements of up to 11.4 per cent in the exchange rate between the Australian dollar and the PNG kina. Secondly, there have been changes in the price of different types of scrap. But thirdly, in the case of aluminium ingots the increase in price of 33.3 per cent is partly a result of more careful selection and smelting of alloys by Hanuacast, giving a higher grade composition. And fourthly, there has been more selection of material to be shipped. For example, in the early days Hanuacast sent scrap car batteries in one shipment, with a price of A$0.15/kg. Subsequently, lead ingots produced after dismantling the batteries commanded a price of A$0.40/kg.

Table 2 also shows that freight costs in the first year represented about 19 per cent of the value of scrap shipped. Better packing and selection reduced this to about 16 per cent in the second year. But an even greater reduction has been possible since August 1984 when Hanuacast, as a regular customer, was able to negotiate a freight charge reduction to a flat rate of K800 per container. Adding in handling charges, the average cost per shipment was reduced to about K870. In relation to recent average scrap values per shipment freight and handling costs should now typically represent only about 11 per cent of scrap value.

The range of types of scrap has also grown through time. This is partly a reflection of how the foundry staff have learned about the relative value of different types of metal (and of how they have passed this knowledge onto the scrap collectors) and partly of greater selectivity in processing the raw scrap. For example, at first, all or most aluminium seems to have been thrown into the reverbatory furnace to produce ingots. By mid 1983, however, it can be seen that some types of aluminium begin to appear separately on the shipment breakdowns. The explanation is clear from the unit prices which show that it makes better financial sense to keep, say, aluminium extrusions worth A$0.90/kg and printers sheets at A$1.00/kg separate from the aluminium alloy ingots worth A$0.80/kg.

In September 1982 the VSO volunteer Peter Thomas arrived in Port Moresby to help in running the foundry. His arrival clearly led to an acceleration in the foundry's

19

development and to some steady improvements in efficiency. For example, the average weight per shipment was 4,740kgs for the first 7 shipments to October 1982: the average for the remainder of the first year's operation was 6,714kgs per shipment. It was at about this time also that the production of lead fishing weights began.

In view of the disappointments of the Lae project it was wholly appropriate that Hanuacast should concentrate on scrap reclamation and that the main work of the foundry should be in melting down aluminium and lead, separating these metals and alloys from steel and other components and thus adding at least some modest value to the scrap before on-selling it. The next stage would obviously be to investigate the possibility of making new products from the reclaimed metals. And it was appropriate, too, to look at lead products first, lead being the easiest metal to melt down and to work.

Lead fishing weights were seen as the obvious product to manufacture. The production process which has been developed by the foundry staff with help from Peter Thomas is described in the next section. Here, the focus is on how the business developed. The idea to make fishing weights was the foundry's initially and the main customer was the Steamships Trading Company, one of PNG's two large retailing networks. But volumes were quite small and the main benefit seems to have been in learning how to organise the production process rather than making significant profits. The breakthrough did not occur until late October 1983.

It was at this stage that Hanuacast was approached by Mr. Brian Kirby, managing director of the Net Shop Pty Ltd. the largest distributor of nets and fishing tackle in the country. Mr. Kirby had forgotten to order replenishment stocks of lead weights from his regular supplier in Japan and without weights the sales of nets (on which the weights are threaded during manufacture) would be affected. Several thousand "medium" weights of 75 grams each were needed urgently.

With encouragement from Mr. Kirby, Hanuacast saw this as an opportunity not just to provide an emergency supply of weights to The Net Shop but perhaps to establish a lasting relationship with a promising local customer. Without really knowing that it would be possible to meet quality and quantity targets, the foundry nevertheless geared up for the effort. On the recommendation of Peter Thomas K500 were laid out to commission a steel die from a local engineering firm. This would be longer-lasting and would produce weights to closer tolerances than the aluminium dies which

The inside of the finished furnace, ready for firing.

21

Hanuacast had used for earlier batches. Within one month the first batch of 600 weights was delivered. Satisfied with the quality and confident that Hanuacast could be a reliable supplier The Net Shop asked for a smaller 37 gram weight to be made. This time, as a mark of goodwill, Mr. Kirby paid the K500 to commission a second steel die. As a quid pro quo he stipulated only that the donated die would not be used to make leads for sale elsewhere. The first batch of 1,430 small weights was delivered in early December 1983. Subsequently, Hanuacast has supplied The Net Shop with various other lead products including split shot sinkers, spoon sinkers and diver's belt weights. Then in February 1985 a large, 150 gram weight was added to the range.

From November 1983 to March 1985 Hanuacast supplied 44,058 small weights, 24,942 medium; and 2,070 of the large size. The total value of these weights to Hanuacast has been K7,263.90. At the current rate of K0.10 per kg paid by the foundry for its supplies of scrap car battery elements, and allowing for a net recovery rate of 75 per cent of usable lead, the costs of raw materials in the delivered weights was only about K500 or about 7 per cent of the value of the weights.

Early Difficulties Encountered

So far, the emerging picture is one of steady learning and progress mixed in with some good luck in the form of the contracts with Non-Ferral and The Net Shop. Crude averages show that Hanuacast pays about K0.178 per kg for raw scrap and sells on to Non-Ferral for about K0.642 per kg — a value added factor of about 3.6 times. Of course, some allowance must be made for net recovery rates of saleable material from raw scrap and for the foundry's labour and overhead costs. Yet fuel is virtually free of charge and capital investment very low. It would seem that the foundry should be highly profitable. But there have been difficulties, mainly financial and management.

An account of the foundry's financial situation will be given later in the report. But as an illustration of Hanuacast's

22

financial difficulties, it is sufficient to say here that in August 1984 it had debts amounting to K19,256. These comprised a bank overdraft of K5,129 and a debt to SPATF of K14,127.

In fact, these financial problems are largely a reflection of management weaknesses, of which there have been several aspects. First, Hanuacast has never been a completely independent entity: it is a trading arm of SPATF. The foundry manager makes day to day decisions about how much scrap to buy, how to process it, when to make shipments, how to divide the labour force between, say, the manufacture of aluminium ingots and the production of lead fishing weights, and so on. But the foundry uses a SPATF-owned truck to collect scrap from the dumps and for other transport needs. It pays a rental charge to SPATF for the use of the truck and for its premises at Hanuatek. Yet it is SPATF that has kept the foundry's accounts files in its own offices, a mile or two from the foundry yard. Hanuacast has its own bank account into which all income from sales is paid. Yet SPATF frequently pays the foundry's wages, debiting the Hanuacast debt account for this, for the transport charges and for other small items. As a result, there is no single person who can see at a glance the foundry's whole financial position. The foundry manager has statements from the bank account, SPATF has the record of the debt account.

Added to the lack of clarity in the accounting procedures, there have been difficulties with the management of the foundry itself. Originally the VSO Peter Thomas was expected to be the full time foundry manager, in parallel training a Papua New Guinean staff member to take over from him. Owing to SPATF staff shortages, however, Peter was asked early on to be responsible for the whole Hanuatek complex. This included a T-Shirt printing business, cane furniture making, car body repair work, a shipping agency, sewing machine repair and various other small-scale enterprises. Coping with a variety of business, financial and personal problems in the somewhat crowded community meant that Peter could spend only a portion of his time with the foundry itself. And it is accepted by all concerned, including SPATF and the foundry work force, that without constant supervision of the right kind, productivity at the foundry declined.

As it happens, if there was a marked decline in productivity and profitability in 1983 and 1984 it is not immediately apparent. Records show reasonably frequent shipments to Non-Ferral with quite high net weights and prices per shipment. The development of the highly profitable lead weight business outlined above also took place in 1983/84.

A full and detailed audit would be necessary to pinpoint exactly how Hanuacast accumulated its mid-1984 debt, and it is doubtful whether the records are sufficiently comprehensive to allow this to be done. Meanwhile, the consensus view includes the following main problems:

(i) lack of close supervision of the foundry work force (as mentioned);

(ii) overstaffing (with extra labour recruited on the basis of the apparently healthy financial position revealed by the bank account, without taking into account the SPATF debt account);

(iii) inaccuracies in weighing scrap at the dumps and overpayment of the scrap collectors;

(iv) double-purchasing of some scrap (which could easily be stolen from the unprotected foundry stock yard).

It is obviously disappointing that by mid-1984 after more than two years spectacular growth in throughput Hanuacast had become burdened with a substantial debt. Yet neither this, nor the problems which caused it, indicate any fundamental lack of viability of the foundry operation.

Current Operations

Hanuacast's current operations fall into five main categories:

(i) selecting, sorting and batching various types of scrap for direct shipment to Australia;

(ii) melting down selected aluminium alloys to form ingots for shipment to Australia;

(iii) manufacturing cast aluminium products for sale locally;

(iv) melting down scrap car battery elements to form ingots

of lead;

(v) manufacturing lead products for sale locally and in Australia.

Scrap Sorting and Batching

This is the simplest of the current foundry operations. Individual scrap collectors at the two Port Moresby dumps or scouring building sites throughout the city assemble their own batches of scrap for sale to Hanuacast, usually once a week. These individual collections can obviously comprise a mix of different types of metal. Hanuacast has by now trained the collectors to make separate piles in six categories according to the price offered. These categories and the rates paid per kg are:

(i) copper and brass together at K0.25/kg
(ii) aluminium castings and extrusions at K0.15/kg
(iii) aluminium foil, cans and gauze at K0.05/kg
(iv) lead (including battery elements) at K0.10/kg
(v) film and film negatives at K0.50/kg
(vi) stainless steel at K0.10/kg

However, there is a much more extensive price structure for different types of scrap offered by Non-Ferral. The arrangement is that Hanuacast can benefit from, say, the A$0.15 price differential between mixed brass and heavy brass only if the sorting of these types is done before shipment. Otherwise Non-Ferral would pay at the lower rate. Similarly, different prices are offered for domestic copper, No.1 clear copper, copper wire, for aluminium extrusions, printers sheets, domestic aluminium and so on. Thus, when the mixed scrap has been delivered to the foundry stock-yard supplementary sorting must be done.

Copper, brass, stainless steel and other metals which cannot be melted down are sorted, with most being packed and compressed as much as possible into 200 litre oil drums. The main exception is extruded aluminium (from window frames, cladding panels and the like) which is usually bulky but not worthwhile cutting down and is packed loose into each seafreight container. Other types of aluminium —

usually castings from motor vehicles, domestic appliances and the like — are put on one side for melting down.

Aluminium Ingot Production

Typical items set aside for the reverbatory furnace are alloy clutch housings and gearboxes from scrapped vehicles, electric motor housings and the like. The rationale for melting them down is twofold. First, many such components still have other types of metal attached to them, including, for example, bushes and bearings, steel studs and nuts and so on. The very first recorded shipment to Non-Ferral contained some aluminium ingots (at a price then of A$0.67 per kg) and some aluminium castings (for which only A$0.50 was offered). These price differences reflect the extra work that Non-Ferral would have to carry out itself to separate the aluminium in the casting from contaminants represented by any fillings of different types of metal attached to the castings. Secondly, by melting down bulky castings and casting ingots a significantly greater weight of metal can be packed into the shipping containers. Freight and handling charges as a proportion of the value of scrap can then be kept to a minimum.

At the same time, Hanuacast cannot afford to spend a great deal of money on converting scrap aluminium into ingots. The structure of mark-ups summarised from earlier figures is as follows:

	Price	Factor
i. raw scrap purchased from dumps	K0.15/kg	—
ii. raw scrap onsold directly to Australia	K0.47/kg	3.13[1]
iii. scrap converted to ingots	K0.63/kg	1.34[2]

1. Estimate derived by applying pro-rata increase of ingot prices 1982 to 1984 to the 1982 price for castings and converted from A$ to kina at December 1984 exchange rate (i.e. $0.8\%/0.67 \times 0.5 \div 1.27$).
2. Conversion of December 1984 A$ price for ingots by current exchange rate (ie. $0.8 \div 1.27$).

In other words, in purchasing, sorting, packing and shipping raw scrap, Hanuacast can increase its value in kina by over 200 per cent. By melting the scrap aluminium and making ingots the foundry operation per se adds only a further 34 per cent to its value. Bearing in mind that the removal of other metals as a result of the melting process reduces the weight by a variable amount, the added value in practice is lower than 34 per cent. More specifically, using prices (ii) and (iii) above, if the net weight of aluminium in ingot form is less than 75 per cent of the weight of scrap put into the furnace no value is added to the raw scrap. The labour and handling costs incurred by the melting and casting operation would then imply a net loss to the foundry.

The melting operation is much the same as originally set up by Willie Feinberg. The cut-away drawing Fig. 1 shows the internal platform where components which do not melt are deposited and can be raked out after each batch of scrap has been processed. Some improvements have been introduced by the foundry staff, however. First, a redesigned double mould has been developed which makes the removal of ingots easier. A set of steel rollers has also been installed immediately below the furnace outlet. This enables the hot moulds to be handled more easily, hence increasing the rate at which new ingots can be cast. A shallow pit has been dug beneath the rollers where leaking oil from the burner and other debris can collect without obstructing the furnace outlet.

One part of the furnace which is less than satisfactory is the burner mechanism. The type of blower shown in Feinberg's IT booklet is no longer being used.[a] Instead a smaller blower of much lighter construction has been substituted. This (Hitachi) blower is not sufficiently robust and there have been frequent breakdowns. Though this type of blower is easily obtainable in Port Moresby and easily connected to the burner mechanism it is far from ideal and the need for constant replacements is almost certainly not cost-effective. The fuel feed control is also not satisfactory — at least the combination of fuel control and the inadequate blower makes it difficult to achieve and to maintain proper atomization of the oil fuel.

a. *Lost Wax Casting: A Practitioners Manual*, W. Feinberg, IT Publications Ltd. 1983.

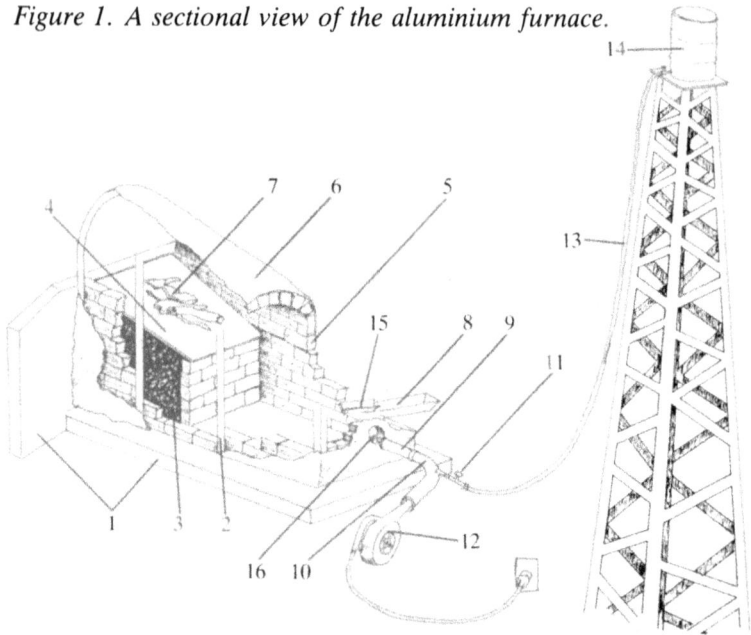

Figure 1. A sectional view of the aluminium furnace.

1. Foundations	9. Nozzle
2. Steel Brace	10. Galvanized Pipe
3. Earth	11. Tap (oil)
4. Steel plate	12. Electric blower
5. Fire bricks	13. Hose
6. Clay	14. Oil drum
7. Aluminium scrap	15. Outlet for melted aluminium and trough
8. Ingot mould	16. Hole for oil to be sprayed into furnace

It has not been possible to measure accurately the maximum temperature which can be reached by the reverbatory furnace. The melting points of the different types of aluminium alloy in the typical mix of scrap are higher than the 660°C required for pure aluminium and the furnace is capable of melting most of these alloys. The best estimate is

28

that temperatures of up to 900°C can be reached but in practice the limiting factor is probably the blower/burner system which needs — but does not often get — constant adjustment to maintain proper atomization and complete combustion of the oil.

On a more positive note, Non-Ferral report that the quality of Hanuacast ingots is now quite satisfactory — better, in fact, than ingots produced in some Non-Ferral branches on mainland Australia. From time to time if particular batches of ingots are contaminated with zinc or other elements to an unsatisfactory degree Non-Ferral inform the foundry and give advice on how to overcome the problem. Usually this is a matter of more careful selection of what to throw into the furnace.

Aluminium ingots awaiting despatch.

Hanuacast ingots and other raw scrap are said by Non-Ferral to represent no more than 0.5 per cent of the company's typical annual throughput. It is encouraging, therefore, that the evidence of correspondence on file at Hanuacast shows that Non-Ferral have been so generous with technical advice and support. Nevertheless, there is no question of any financial subsidies or premiums having been paid by Non-Ferral. The prices they offer to Hanuacast are strictly commercial, based on metal content analysis and prevailing market rates.

Once the Hanuacast ingots have arrived and been analysed by Non-Ferral they are blended into new alloy mixes to international standards. (This is one reason that Non-Ferral offer a higher price for the purer types of aluminium in extrusions and printers sheets as it enables them to adjust alloy blends at their own foundry.) After blending and refining, about 50 per cent of "new" alloy ingots from Non-Ferral are on-sold to the motor industry where they are used in making — once again — clutch housings, gearboxes and so on. Other products requiring perhaps different alloys include street lamp and other electrical fittings, motor mower components and barbecue grilles.

Manufacturing Cast Aluminium Products

Hanuacast's own ventures into making new cast products for sale locally are still at a very early stage. There is no doubt that in simple terms of potential revenue per kilogram and value added to raw scrap, the making of new products ought to be an interesting prospect. There are, however, four main factors that the foundry knows must be taken into account, all of which are inter-related:
  (i) identifying products and marketing outlets;
 (ii) refining casting techniques to achieve the required quality;
(iii) labour costs;
(iv) foundry layout.

In product identification it is not just a question of looking at what cast aluminium products are currently imported to PNG and which of them could conceivably be made with the techniques and equipment available to the foundry. It is also a matter of whether alloy composition is important (because Hanuacast has only rudimentary control over this); dimensional tolerances in the basic casting; and quality and appearance of the finished product. In addition, heterogenous products with a variety of markets could mean that considerable work would be needed in developing contacts with various marketing and distribution organizations.

The basic casting techniques that would be involved have been described in Willie Feinberg's IT booklet. Hanuacast has already found that silica sand (not coral sand) is available locally, though it is not as fine as would be used in a modern foundry and therefore does not give such a good finish to the casting. Clay, too, is available of the right type to give a good coating to the sand while also standing up to the heat from the molten aluminium. But instead of graphite or very fine coal dust (which is used to coat the inside of the mould and produce a cushioning layer of gas when it is burnt off by the molten metal) Hanuacast has been experimenting with crushed charcoal which is not quite so effective. Thus, again depending on the quality of finish required in the final product, more work, more time and more money may have to be spent to produce satisfactory results.

The manufacture of cast products in mould boxes in any case requires much greater labour inputs per kilogram of aluminium than the straightforward process of rough casting ingots in open steel moulds. And to a certain extent deficiencies in sand, clay or charcoal in the mould could be reduced by still more labour inputs in filing, grinding, polishing and the like. But these labour inputs would have to be carefully costed to ensure that the operation is still profitable.

If cast products were to become a significant part of the foundry operations it would also be important to ensure that the foundry layout lent itself to efficient work processing. The Hanuatek foundry yard had clearly not been laid out with large-scale product casting in mind. As a result, when

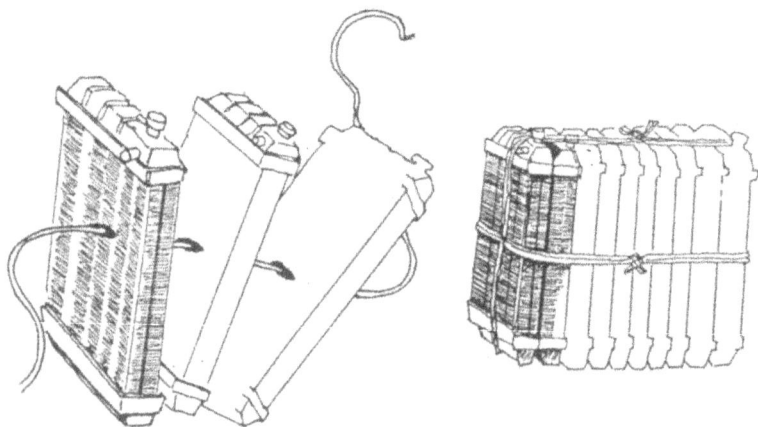

Figure 2. Showing how old radiators are threaded together prior to shipping.

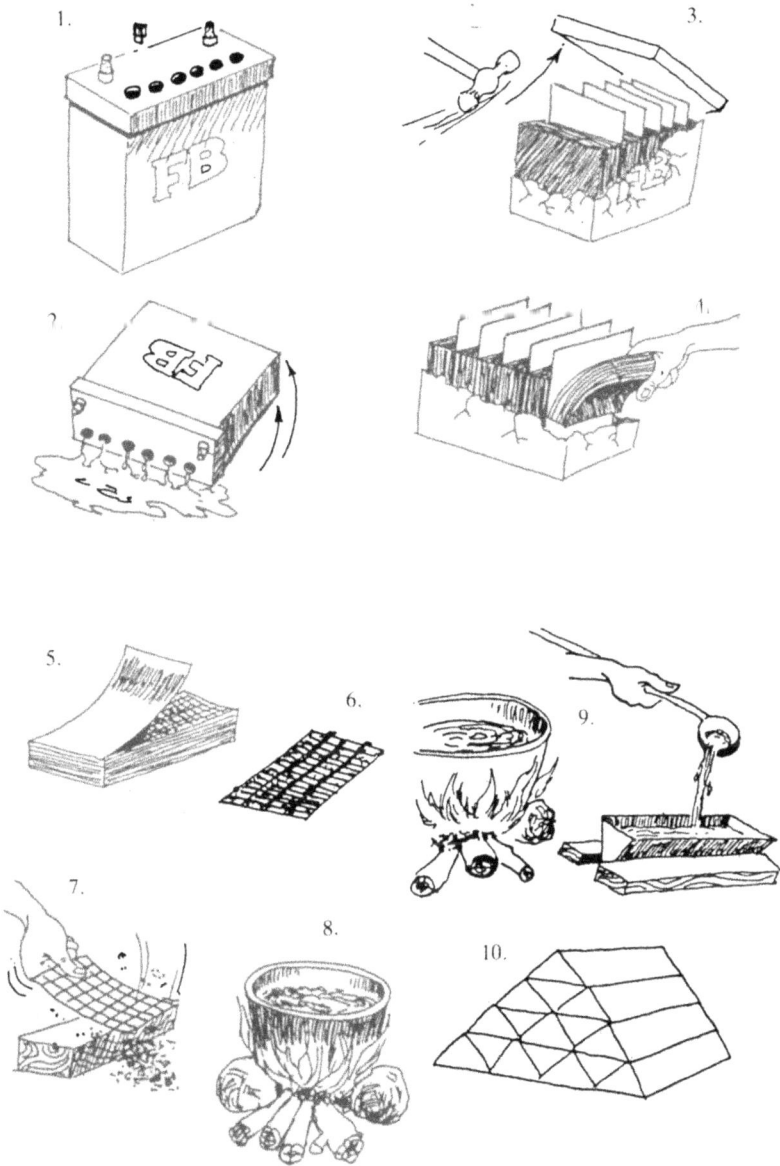

Figure 4. Old car batteries are melted down and made into lead ingots.

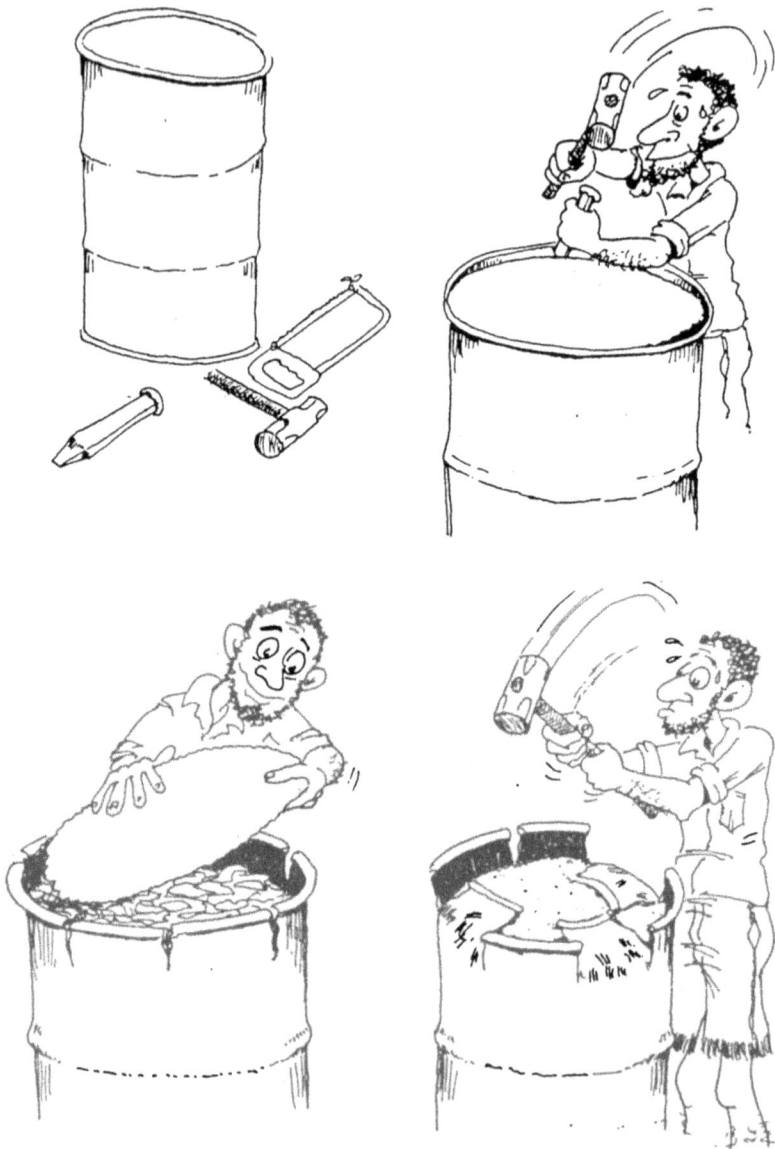

Figure 3. Illustration showing the "drumming" of scrap for shipment.

products were being made it was difficult to maintain normal rates of ingot casting. Indeed, not only would more space be necessary but probably a separate furnace would have to be set aside specifically for product casting, away from the debris generated by raw scrap processing operations, and laid out for rapid casting of batches of products.

It is only recently that the foundry has attempted the commercial manufacture of new cast products. The main item is an aluminium bracket for the Electricity Commission (Elcom). This is designed to be clamped at the top of transmission poles and to hold the horizontal cross-members from which the cables are suspended. It has not been possible to look in detail at the structure of production costs for this item, of which several batches have already been delivered. (To do so would involve a time and motion study to separate out labour costs in the manufacture of this product from those in the parallel operation of making simple ingots.) But in relation to the four factors mentioned above, the experience has been useful. First, the casting is relatively simple and there is no marketing difficulty since Elcom's central office is the sole customer. Secondly, as a rough, outdoor fitting Elcom is concerned only that the bracket should fit the transmission poles, should accommodate the cross members and should be strong and durable. Alloy composition is not critical. Quality of finish is also not a prime consideration. But thirdly, it seems likely that Hanuacast is spending much more in labour costs on this product than necessary, thus eroding its profit margin. One reason is that because this is the first commercial product and because the foundry workers are proud to have been given the order by Elcom they tend to spend more time than necessary in grinding rough edges and tidying up the casting. Another reason for likely high labour costs relates to the fourth factor — foundry layout. Four or five moulds have been made for the cast brackets, but the yard is too congested for rapid batch casting. And since foundry workers are on flat daily wages the longer it takes to prepare, pour and strike each casting, the higher the labour cost in its production.

36

Melting Down Battery Elements

This is another operation for which the technique and process have been developed by the foundry staff themselves after the first visit by Willie Feinberg. It seems to have been originally envisaged that the vertical oil fired drum furnace would be used for melting down lead battery elements. But the capacity of a crucible in such a furnace is quite limited and in any case the low (330°C) melting point of lead did not justify using such a furnace which is capable of much higher temperature. The alternative developed at the foundry is a large, shallow, open crucible supported by bricks and clay about 50cms above a vented hearth with timber as the fuel. The crucible, of steel, is about 1.5-2.0 metres in diameter and about 50cms deep and originally formed the end plate of a large boiler. The timber fuel, like the oil used at the foundry, is free of charge from a local sawmill. The advantage of this system is that it can take larger quantities of battery elements and the large surface area of molten lead allows scum and impurities to be raked away. The minor disadvantage is that the shallow crucible has no outlet. Molten lead has to be ladled out and poured into moulds forming ingots either for sale as ship's ballast or for further processing at the foundry.

Lead Product Manufacture

The history of Hanuacast's involvement in making lead weights has already been outlined. The technique is quite simple. For this operation the vertical oil-fired drum furnace is used since it is capable of holding sufficient lead for the rate at which the weights can be cast. The raw material is the lead which has already been melted once in the large crucible. But the drum furnace has been modified from the Feinberg design. Instead of a removable lid with vent hole the drum has been left open to accommodate a crucible which is usually a rear axle differential housing from a truck or similar vehicle. This is laid across the top of the drum and packed round with bricks allowing small vent gaps as necessary. This probably reduces the furnace temperature below the maximum it could reach with a proper lid. But again because of the low melting point of the lead this is not a problem. The main deficiency is that the lead, even after a first melt, still contains some sulphuric acid from the car batteries and this acid erodes the steel in the crucible. Each differential housing lasts only for a few weeks and Hanuacast is finding that supplies of scrap differentials have become more difficult to find. Properly designed, thicker crucibles would have a longer life — but would cost more.

As mentioned, the dies currently used for Net Shop orders are made of mild steel, giving greater durability and a better finish than aluminium moulds (which could otherwise be made by Hanuacast itself). They are of the split type, clamped together during the casting process and most dies now used are designed with chambers to produce three fishing net weights on each pouring. Removable steel pins through the centre of each chamber produce a hole through the furnished cylindrical weight, enabling it to be threaded onto the net.

The interconnected chambers and central pouring hole in these dies mean that each set of three weights needs some trimming work to separate them and remove webs and lips of lead. But once cool the lead is still so soft as to make trimming an easy and rapid process. Lead trimmings are then thrown back into the crucible.

Safety Procedures

Safety procedures throughout the foundry yard are at best rudimentary and SPATF has recently drawn management's attention to the deficiencies. Foundries are always dirty and potentially dangerous places but some basic precautions would seem to be quite necessary. One example is the reverbatory furnace. The quite large mouth of the furnace is necessary to enable large pieces of scrap to be loaded in. But during firing there can be small explosions (from, say, blocked cavities in an aluminium casting) and pieces of scrap can "backfire" out of the mouth where a foundry worker is likely to be standing. What is needed is a protective grille over the mouth to catch any flying pieces of metal.

Secondly, the foundry staff have shown little inclination so far to wear proper protective clothing or footwear — not least perhaps because a foundry in a hot climate can be a very hot place indeed. But this does mean that there are risks of serious injury from molten metal. Thirdly, the yard's electrical system feeding the furnace blowers, overhead lights, grinding machines and so on is best "jury rigged" and at any time, especially during wet weather, there do seem to be risks of short circuits.

Some of the safety deficiencies can be seen as a result of the gradual development of foundry operations and work-processing systems. If, for example, it is beneficial to move a grinding operation from one bench to another or to move the position of a furnace blower as an experiment there is a tendency to make temporary adjustments to cable runs. But if such an ad hoc arrangement is not then properly revised and laid out it can become a source of danger. This all suggests that there should be periodic reviews of these temporary arrangements during which they are tidied up and proper working procedures are developed. In the case of Hanuacast it is to be hoped that the move to a new site will provide an opportunity to improve physical layout and installations in the interests both of safety and efficiency.

As part of the safety procedures for the foundry, a monitoring system to test the blood lead levels of foundry workers has been started. The first tests on three workers showed that two had readings of 2.4 umol/L and 3.4 umol/L respectively, which is above the normal range of 0.0-2 umol/L, but below the level of 4.5 umol/L at which lead poisoning is considered to begin.

The readings were taken when the two workers were involved with lead at the Hanuatek site. At the new site (see Appendix II), conditions are much better and it is expected that lead contamination will be lower. The conclusion has been drawn that there is no need for SPATF to close the lead section, but that safety precautions such as showers, filing under water, face masks, washing down the site, etc., should be adhered to rigorously.

Management and Staffing

The lack of clarity — at least the split — in management functions between the foundry and SPATF has already been mentioned as one difficulty in the development of Hanuacast since 1982. But the split is still there and raises a number of issues not only to do with management capacity at the foundry but with SPATF's policy for the role and future of its subsidiary, Hanuacast, and of the other small enterprises in the Hanuatek complex. There are issues which will be dealt with later in the context of the outlook for the foundry.

Looking first at the foundry itself, it is clear that management capacity is still insufficient for the complete running of all aspects of work. Four aspects of management in particular can be singled out to illustrate the issues — labour supervision, financial control, technical development and planning.

Labour Supervision

There have been — and still are — times when productivity in the various foundry operations has slipped. Of course, if a furnace blower breaks down it must be repaired or replaced. If a truck differential crucible begins to leak molten lead something needs to be done about it. But problems of these types can mean that several of the foundry's nine yard workers will become involved, affecting productivity in the other operations. Or the blower/burner mechanism on the reverbatory furnace may work loose, giving incomplete combustion, yet the worker on furnace duty may not notice the lack of adjustment. Or by mistake the wrong type of scrap might be thrown into the reverbatory furnace, producing a contaminated batch of aluminium alloy ingots. These and similar problems mean that the yard workers need quite close supervision if quality and throughput are to be maintained.

Part of the problem is perhaps related to the generalized notion of Papua New Guinea's inexperienced industrial labour force, as mentioned in Section II of this report. In the particular case of Hanuacast, however, it must be remembered that for all concerned the foundry is still a relatively new venture, that the experimental nature of the project which itself generated so much interest and excitement is still present in people's minds. And although in one respect the foundry is a flourishing business with customers waiting for deliveries and so on, the sense of urgency that these external realities ought to create has not yet been passed on to the work force.

Financial Control

Although, as mentioned, SPATF headquarters operates a Hanuacast debt account and keeps detailed data on truck hire charges, wages and other aspects of foundry operations, the foundry office does keep some other financial records. The principal types of financial data in the yard office are weekly payment rosters to the scrap collectors; price and quality breakdowns of scrap shipments which are returned to Hanuacast after payment by Non-Ferral; details of local sales of lead weights, ships ballast and other products; and statements of the foundry bank account (into which Non-Ferral payments are made).

It is not just that it would be neater and more in line with normal practice for the foundry management (whoever it is) to have ready access to all financial data. More substantively, it is only possible to have financial control over operations if management can analyse what costs are being incurred, for what operations, and with what returns to expenditure. Operational decisions about the types of work to be done, about staff recruitment and disposition and so on depend centrally on financial information. In the past, SPATF has felt that the foundry might be overstaffed. But if foundry-level data suggest that an extra worker can be afforded it is not surprising that he is recruited. On the other hand, SPATF might have good financial reasons against new recruitment yet not be sufficiently familiar with foundry operations to appreciate how an extra worker might facilitate a particular operation and lead to extra profits fully justifying the marginal increase in wage costs.

In summary, the lack of financial control is the fault neither of the foundry nor of SPATF but of the split in record-keeping and in authority.

Technical Development

In spite of the very encouraging development the foundry has achieved in its relatively brief history much still remains that should — or could — be done. One example, is the burner/blower mechanisms on the oil-fired furnaces. Greater reliability and control would make a significant contribution to efficiency. And cleanliness could be improved if oil leakages from around the burner nozzles were reduced. As it happens, a large electronic motor and fan has already been installed near the reverbatory furnace but it has never been connected because further work is needed in making air ducts and adapting the burner assembly — work which nobody has had time to do. Another example is development work with new moulds, the search for finer grades of silica sand, better quality and better pulverization of charcoal for moulding work and the like.

Technical development work of this sort may not be part of day to day management but it is an essential input to management decisions. It is the type of work which ideally could be done by an expatriate volunteer. Beyond this, the Hanuacast yard has so far been much too small to accommodate any experimental work without affecting normal operations. What is needed is a specific area set aside for technical development with its own furnace or furnaces in which new techniques can be tested.

That is not to say that experimental work is not being done. Indeed, it is a measure of the personal interest and involvement of the foundry staff that if a new experiment is being carried out all idea of normal working hours is forgotten and the workers will stay behind to see if the experiment can be made to succeed. An illustration of this took place in March this year. A government agency had expressed interest in buying a large but simple brass casting about 50cms square by 15cm deep. Work on lead weight production stopped and an effort was made to melt down some brass in the vertical drum furnace. By 6.00 p.m. the brass had not quite melted and the experiment looked to be disappointing. But the half dozen workers all seemed to be reluctant to give up and go home. In other words the will to explore new possibilities is firmly there; it needs only the right environment and organization.

Planning

This is an aspect of management which pulls together all the previous three. Given a particular set of activities (eg. sorting and selecting brass, copper and non-meltable metals, aluminium ingot production, lead weight manufacture) it is only if labour productivity is known, if the costs and returns in each activity can be isolated, and if techniques have been fully developed that it becomes possible to plan what emphasis to give to what type of activity. At present Hanuacast is barely able to do this with any confidence. It is involved in its current activities because the opportunities presented themselves and prima facie seemed profitable. There is nobody either in the foundry or in SPATF who currently could tell how extra emphasis in one activity which might have a high profit margin might affect (positively or negatively) the overall profitability of the entire foundry. Questions of this sort have begun to be asked by SPATF but much more information would have to be collected in the foundry yard itself before they could be answered.

Planning — based on a thorough analytical knowledge of existing operations — is also important for possible new activities and products. Having established a few basic techniques and procedures the way forward for Hanuacast in the near term would seem to be (a) to ensure the best possible supplies of scrap, and (b) to add more value to the scrap in the foundry by manufacturing more products. But what sorts of products — in what quantities — and selling for what prices? These questions are not easily answered. But if possible new products can be identified another set of questions is generated in the planning process. Should a given new product be made in addition to maintaining the scale of current operations (in which case more labour and perhaps more space would be needed) — or should there be a reduction in some current operations? This is the familiar issue in project planning of assessing the viability of a new venture both in terms of absolute returns and in relation to other possible choices. At present the foundry and SPATF are both ill-equipped to make such assessments.

Current Foundry Management

Over the past year the VSO Peter Thomas has been able to reduce the amount of time spent with the other Hanuatek entrepreneurs and focus more on Hanuacast itself. Another equally promising development was the semi-official appointment of one of the foundry staff, Mr. Wilson Amos, as trainee manager.

Wilson's background is in some ways a paradigm of what confronts an ambitious young Papua New Guinean looking for an opportunity for a rewarding occupation and keen to take on responsibility. He left school in the middle of Form 2 (the second year of secondary education) and at the age of about 18 joined the Defence Force as a storeman. Looking for better prospects, he left after 18 months and joined a large pharmaceutical retailer, again as a storeman. After a while he was given an opportunity to take part in a 3-month sales training course. But in 1981 he left the pharmaceutical company, disagreeing with the way the sales operations were being managed. There followed a period of about 6 months when he was unemployed in his home village, a few miles from the centre of Port Moresby. He managed to find some work as a driver on contract until he learned of Hanuacast through his brother who was working there. The foundry took him on in early 1983 as a casual worker breaking up old car batteries for 6 weeks at K20 per week. Then he was taken on full time in processing lead and he was involved in this when the first Net Shop order came in and lead weight production suddenly became a commercial operation. Largely as a result of his reliability in this work he was made semi-official trainee manager in early 1984 and 6 months later he took over from Peter Thomas as the official manager. He is still less than 30 years old.

Wilson is quite open about what he would like to learn — and quite accurate about his priorities. One area in which he would like to increase his knowledge is in the technical aspects of foundry work, casting techniques, the mixing of different types of aluminium alloys for different purposes, and the organization of production at the foundry yard level. In these respects he would like to visit other foundries to study their techniques and organization. But even more, he would like to learn business management. Commenting on the arrangement whereby SPATF has been keeping some financial data at its headquarters office, Wilson approved of this at first when he became Manager because it was as much as he could do to establish his authority with the work force and to keep production going. Since then, however, he has come to realize that he needs access to full information about foundry finances and cash flow and has reached an agreement that SPATF will return all the files to the foundry office.

Labour Force

The foundry's total staff complement in March 1985 was 13, comprising the following structure and wage rates.

	Kina/week (flat rate)
Manager (Wilson Amos)	75.00
Management Adviser (Peter Thomas)	67.50
Book Keeper	67.50
Yard labour force (9 men @ 32.83*)	295.47
Driver	28.80
Total weekly wage bill	534.27

*(average figure: range is from K55.00 to K15.00).

Some of the foundry yard workers have been employed

since the earliest days — and these tend to be the highest paid. The lower wage rates are paid to young men recently taken on and in a sense still in a probationary-cum-training period. The terms of employment are the normal 5-day week and if overtime is worked payment is on the same flat rate as the hourly equivalent of the weekly wage. Seven of the foundry yard work force and the driver are paid less than the legal minimum wage for Port Moresby of K45.20 per week. But the foundry is by no means the only employer in the city paying low wage rates and it seems that the authorities do not enforce the legal minimum with any great vigour.

The Scrap Supply System

Each of the two main rubbish dumps for Port Moresby, at Baruni on the 'back' road and at Six-Mile near the airport, is associated with a nearby group of urban migrants or "squatters" many of whom have added scrap metal collection to the range of informal sector activities outlined earlier in this report. In addition, there is a third, larger group of "wandering" scrap collectors who range throughout the city looking for items of value to Hanuacast.

The Baruni dump is more isolated than Six-Mile and seems to have a smaller squatter settlement and smaller numbers of scrap collectors. The Baruni settlement is also more ethnically homogeneous. Most settlers come from the Goilala area, north west of Port Moresby. Interviews with six scrap collectors revealed that four had joined the settlement three years ago while two had been living there for more than ten years. All described themselves as "wantoks" of the rest of the group, and all were living there with their immediate families. These who had been there the longest said that scrap collection is not a new activity for them. Long before Hanuacast came into being they had been able to sell scrap car batteries and some other items to a German entrepreneur. But in the last two or three years Hanuacast had become a more regular and reliable customer, bought a wider range of scrap, and paid better rates. As a result, scrap collections had come to figure more prominently in their total

46

income. One man said that he and his wife could earn between them K25-50 per week from selling scrap.

None of the Baruni families appear to rely wholly on selling scrap. Their home area is well known for the fertility of its soil and although the entire Port Moresby area is poorly suited to agriculture the Baruni settlers all have their own gardens. Typical crops include tapioca, yams, corn and bananas, and peanuts which are sold in local markets. This pattern bears out the research conclusions that the core of much informal sector activity in PNG is urban gardening. Collecting and selling empty bottles was another activity quoted as bringing in a little money. None of the scrap collectors interviewed at the dump had ever been able to find a job since coming to Port Moresby. None had any special skill.

The Baruni settlers have built their houses on a small knoll, surrounded by their gardens, about half a mile from the dump itself. Many of the building materials used were salvaged from the dump. At one time some of the scrap collectors had built small houses or shelters actually on the dump — but the health authorities had instructed the police to remove them. At the main settlement none of the squatters is obliged to pay regular rent to the land-owner. Sometimes, however, if the land-lord has to assemble a bride-price or has some other financial obligation he approaches the settlers to make a contribution.

Some of the Baruni scrap collectors said that they did sometimes consider going back to their home village. When this might be seemed to hinge on the amount of money they might be able to save. The underlying feeling is that to go back empty handed after a period in the city would reflect badly on their status in the village.

The Six-Mile dump is surrounded by a much larger number of settlers of more diverse origins. Still, individual clusters of houses tend to belong to groups of people with similar areas of origin. Some contain settlers from Goilala, the same group as at Baruni. Others are from the Highlands, particularly the overpopulated province of Simbu. And in these groups there is a higher proportion of single men. (The polygamous customs in parts of the Highlands mean that there are many young men who cannot find wives and some of them come to Port Moresby looking for new opportunities.) These single men tend to be less involved in gardening and more reliant on scrap collecting and other ways of earning money. But at the same time they have only themselves to support and can better afford to buy food.

Surrounded as it is by more squatter settlements there is much more competition among scrap collectors at Six-Mile than at Baruni. Trucks entering the dump are immediately surrounded by 40 or 50 people, including young children. The more athletic jump onto each truck even before it has stopped, beginning the search for anything that might be of value. While this probably means that scrap collection is quite efficient the risk of injury among the collectors is high. Some of the Six-Mile squatters who have been living on or near the dump for as long as 20-25 years say that it has become distinctly more difficult for individuals to make a reasonable amount of money from scrap owing to the competition. At the same time, none of the five collectors interviewed relied wholly on selling scrap. Nor is there any evidence of the longer established settlers attempting to limit the numbers of people competing for the scrap.

For scrap collected at each of the dumps Hanuacast sends its 2.5 tonne truck — usually about once a week — to weigh the different types of metal and carry it back to the foundry. Payment, however, is not made during the pick-up. Instead each person is given a small docket indicating the type and weight of metal. Then, one day a week is set aside in the foundry office when docket holders can come to be paid.

The "wanderers" bring the scrap they have collected direct to the foundry but they too are given dockets for later payment. This system enables the foundry book-keeper to calculate from carbon copies of the dockets the exact amount required for each weekly payment session. An appropriate withdrawal of cash from the bank is made the day before so that large amounts of cash do not have to be kept on the premises longer than necessary.

Weekly payment rosters are kept at the foundry office

with each collector asked to sign or make a personal mark against each payment for each type of scrap. In each weekly roster there are many payments of only a fraction of one kina for just a few kilos of metal. In some weeks, however, some individuals (often the same ones from week to week) sell much larger amounts, with earnings of up to K60-80 per week.

Unfortunately the weekly payment rosters do not show how many individuals from the same family sell scrap each week. On payment days there is a mixture of men, women and children queueing up with their dockets. Some may be from the same family; the children may or may not give their parents the money they earn. But even assuming that all of the Baruni group were from different families, and bearing in mind that the Baruni settlers can also sell produce from their gardens, their weekly earnings from scrap represent about 50 per cent of the legal minimum wage and about 70 per cent of the average weekly wages of Hanuacast's yard workers.

Another reference against which the average earnings from scrap can be compared is the informal sector research study quoted above. From this it can be seen that the average earnings from scrap for Six-Mile, Baruni and wanderers represent 54 per cent, 91 per cent and 36 per cent respectively of the average weekly *household* incomes in the Gordons Ridge settlement among households without a wage earner. It would seem, therefore, that for households which do have an opportunity to collect scrap their earnings from it can be a major contribution to total household income.

Based on 1980 census data some very crude estimates can be made of the effects of Hanuacast's scrap purchasing operations in relation to the total number of urban households without a wage earner in and around Port Moresby (an area known as the National Capital District, NCD). In 1980 there were about 1,120 such households in the NCD, representing over 14 per cent of the national total of 7,740 non-wage-earner urban households. Since then, the Institute of Applied Social and Economic Research has projected a gross population growth rate of 4 per cent per annum for the NCD, nearly double that for the country as a whole. Crudely applying this growth to the non-wage-earner urban households suggest that in 1984 the total would have risen from 1,120 to about 1,300. Looking back at the Hanuacast average weekly roster of 133 separate payees, income from scrap collection could be reaching a maximum of about 10 per cent of non-wage-earner households — assuming that each payee represents a separate household. More realistically, it is possible that Hanuacast's payments are making an income contribution to anywhere between 5 and 8 per cent of non-wage-earner households in the NCD. For a single small-scale enterprise this represents a most impressive spread of benefits.

Financial Aspects

It has been beyond the scope of this study to undertake a thorough financial analysis of the Hanuacast operation. The split in accounting records between the foundry office and SPATF referred to earlier is one complicating factor. Another is that the variable disposition of the yard work force among the different main operations (e.g. sorting and packing; lead and lead products; aluminium ingots and products etc.) makes it difficult to assign discrete costs to each. Sample observations over a longer period of time would be necessary to build up an accurate picture of cost centres and the structure of production costs.

Nevertheless, some financial information has been directly gathered, and SPATF itself has since mid-1984 shown an increasing concern about Hanuacast's financial health. And from these two sources some broad indications can be assembled on the structure of foundry costs and revenue.

Income

The foundry's total income is dominated by sales to Non-Ferral. Other sources are the sale of lead ballast and fishing weights and much smaller amounts from sales of other products such as the cast aluminium brackets for the

Electricity Commission.

Analysis of the most recent 6 shipments to Non-Ferral (from mid-August 1984 to the end of November, 1984) for which data was available on actual payments shows an average value per shipment of about K7900. (This figure is based on exchange rates prevailing on the date of preparation of the bill of lading, not when payment in Australian dollars was actually made.) During calendar 1984 a total of 15 shipments were sent, giving a derived total value of K118,300.

Inspection of the foundry office invoices for sales of lead and lead products showed a sub-total of K8,047.7 for the period 14th February to 21st September inclusive. On this basis a projected annual income of about K13,350 could be expected.

No data could be assembled on the value of other sales but a realistic estimate might be at about 25 per cent of the value of lead sales — or K3,300 per annum.

Using these estimates, Hanuacast's total annual income can be put at about K135,000.

Expenditure

The largest item of expenditure is for purchases of raw scrap. Inspection of a sample of scrap reconciliation records for nine regular weekly payment days over the period 4th October 1984 to 14th February 1985 showed an average weekly cost of K1,401. Assuming a full year of 52 weeks, the total cost of scrap procurement would be K72,852.

The next largest cost item is labour. For costing purposes the weekly wage bill of K534.30 can be reduced to K466.80 by deducting the K67.50 income of Peter Thomas. But an allowance of about 10 per cent can be added in to cover average overtime payments. Again assuming a 52-week year, the annual wage bill would be K26,700.

Other cost items are site rent of K32 per month (payable to SPATF); transport rental estimated at about K600 per month; electricity at K25 per month and miscellaneous items at about K870 per shipment and assuming 15 shipments per annum would typically amount to K10,440 per annum. This, added to the annual cost of the transport, rental and other items above gives a total of about K19,050.

Total annual expenditure would thus amount to about K118,600. In simple terms, therefore, the foundry's excess of income over expenditure would be K16,400, representing profit margins of about 14 per cent on costs and 12 per cent on the value of sales.

SPATF Analysis

In August 1984, SPATF carried out its own financial analysis of Hanuacast covering the first 7 months of the calendar year. The analysis shows a lower income from sales to Non-Ferral (but is based on an earlier sample of shipments than those quoted above). But it shows a higher value of other income than the estimates above. The SPATF estimate of expenditure, however, is very similar. Combining the two sets of estimates gives the following.

	SPATF (August 1984)	This Study (February 1985)
Annual income from shipments	88,284	118,500
Other Income	22,044	16,650
Total Income	110,328	135,150
Total Payments	117,768	118,600
Gross Profit (Loss)	(7,440)	16,550

These two estimates are, of course, critically different. But their differences are hard to reconcile. Both analyses include estimates and in the case of this study some of the estimates are based on grossing up income and expenditure items to an annual basis from sample observations (of, say, weekly payments for scrap). The value of the analyses lies, perhaps, in demonstrating that Hanuacast appears to be working to very close margins. It has to be borne in mind that the foundry's fuel costs are zero except for transport. And whichever of the above two estimates more nearly resembles the current truth, Hanuacast could ill-afford to pay commercial rates for fuel. Looked at in another way, if Hanuacast paid the minimum legal wage to those of its workers currently below K45.2 per week, its flat weekly wage bill (excluding the VSO adviser) would rise from K466.80 to K606.60. Adding the same 10 per cent allowance for overtime would bring an annual total of K34,697.52 — an increase of nearly K8,000 over the estimate. This would reduce the more optimistic profit estimate to K8,550, representing a margin of only about 6 per cent on the value of sales.

On a more positive note, Hanuacast's relationship with its bank, the ANZ Bank in central Port Moresby, is a good one. In common with other banks, ANZ generally has little confidence in indigenous entrepreneurs. It also looks unfavourably on efforts to tamper with strict commercial methods of working and reports that its experience with the government's Credit Guarantee Scheme has been 'disastrous'. It is all the more to Hanuacast's credit, therefore, that ANZ regards the foundry as having established an acceptable track record. At the same time, one important element is that the foundry deliberately chose ANZ because Non-Ferral uses the same bank in Australia and this means that credits in payment for scrap shipments can be quickly processed. Without this link, the foundry may have experienced more difficulties — at least in its early period. ANZ is also unaware of the debt owed by Hanuacast to SPATF and that transactions through the foundry account do not reflect the full picture of its cash flow.

Conclusions and Outlook for the Future

The first, most obvious conclusion is that Hanuacast belongs to that special group of enterprises which are small, based on a simple, cheap technology, and successful. Its success has been based on a number of factors. It has been managed and staffed by an enthusiastic group of people; it developed a workable technology; and it devised a set of operations which have occupied the right economic niche between the supply of scrap on the one hand and the effective demand for processed scrap and products made from scrap on the other. In reaching its present position there have been elements of good fortune, of course. If the relationship with Non-Ferral had not developed so early on, the foundry might have found itself trying to make saleable products for sale domestically as its prime objective — and in this case it would probably have failed. If the Net Shop had not forgotten to place a regular order for lead weights from Japan, the foundry's local sales might still be at a very low level. On the other hand, Hanuacast has faced — and still faces — a number of difficulties. Its margins are very tight. It needs to increase the value added to scrap, it needs to increase its labour productivity, or both. Its ability to play for a greater throughput of scrap using the same labour force must be limited in view of the apparent efficiency of the scrap collection system that has now been achieved. New products and possibly new methods of working may be the answers. Yet neither the foundry nor SPATF yet have the necessary information about the existing composition of costs to be able to assess the financial viability of different alternatives.

In fact it is not possible to assess the future for Hanuacast without looking again at its relationship with SPATF for the foundry is not an independent operation, it is a subsidiary project of the Foundation.

SPATF began as an agency focusing on appropriate technology in and for the rural areas. In the last five years or so it has had an involvement with small-scale urban enterprises, of which Hanuacast is perhaps the best known and most successful. But what does SPATF expect of Hanuacast now? Is the foundry to serve as a model for the original target of setting up similar foundries in Mt. Hagen, other towns and rural villages? This may be possible in the larger towns. But the supplies of scrap and of free fuel oil simply would not be found in a village and any rural foundry or blacksmith operations would have to be based on making products primarily. Once the foundry has been able to clear its debt to SPATF should it be launched as an independent enterprise, a cooperative formed, say, from the existing staff? Or should it be retained by SPATF? If the latter, should financial viability be the foundry's principal indicator of continued success? Or should it spend all of its profits in supporting further experimental, development and training functions tailored to the establishments of new, perhaps even smaller foundries elsewhere in Papua New Guinea?

SPATF is asking these and similar questions of itself and has not yet decided what the answers should be. There are, however, some indications of what it might decide. SPATF is under-funded. It receives core finance from the government, but only on an annual 'project' basis. SPATF also still sees its long-term main aim as helping in rural development. The Foundation needs more staff and it would like to promote its rural extension services which have never achieved the depth of influence or the geographical coverage that had been hoped. If SPATF could contemplate recruiting more staff it would need at least K5-6,000 to provide a house for each of them. Against this background the likelihood is that SPATF would retain Hanuacast, would help its continued development, but would look to the foundry as a source of income to support other SPATF initiatives.

As far as the foundry itself is concerned there are more pressing considerations to face than what SPATF may have in mind. In March 1985 the foundry planned to move to a new location, the Small Industries Centre, which had previously been managed by the Department of Industrial Development. Some opinion in the Department would like to transfer the title to this site directly to SPATF, enabling the Foundation to charge economic rents to the entrepreneurs occupying it and thereby adding to its income. It remains to be seen, however, whether the woodworking, ceramics and footwear enterprises which are already in operation there would be any better placed than Hanuacast itself to afford to pay such rents. Nevertheless, Hanuacast will have more space. It will be able to build more furnaces, to design its layout and services more efficiently, all of which could pay great dividends in terms of productivity.

It is also clear that the foundry manager and the staff

would like to add to the range of cast products in lead but more particularly in aluminium. And this desire stems not just from financial consideration but because those involved feel that it is more interesting to make things of some more direct use than, say, ingots of aluminium.

In the short run the strongest possibility is a new range of lead fishing weights for export to Australia, where there are only about three companies involved in distributing these items. An import/export agent visited Hanuacast on 27 February 1985 to inspect the foundry and assess possibilities. Any order that might come of this contact is likely to be considerably larger than the volume of current business with The Net Shop.

Products in the form of aluminium castings are more difficult to identify. At the foundry there has been some discussion about making cast aluminium patio furniture. This might be possible but such castings would be extremely complex and although a complete patio set might command a price of more than K1,000 (judging from existing prices), customers would expect high standards of finish. More promising might be to look at the example of the transmission pole brackets already being made for Elcom. Hanuacast could perhaps approach the Committee on Government Procurement Procedures and Policies over the policy to promote domestic procurement whenever possible. It could ask for access and introductions to all government agencies simply to inspect what is currently being procured from overseas and to assess what items the foundry might be able itself to produce. If Elcom needs brackets, perhaps the Public Works Department needs similar, simple fittings for road signs — and so on. Such products would not have the glamour of patio furniture, but they might represent a better contribution to national development.

Replicability

Within Papua New Guinea the best prospects for new foundry operations would be, as mentioned, in the larger towns. The two critical factors, judging from Hanuacast's experience, are supplies of scrap and of free fuel. The first is a function mainly of (a) construction work (producing off-cuts of aluminium and copper wire); (b) motor traffic (producing scrap vehicles); and (c) high income urban settlement (producing scrap appliances). The second is also a function of motor and/or marine traffic producing waste oil as fuel.

Careful assessment would be necessary, however, before it could be shown to be viable to set up similar scrap recovery operations. For example, it is likely that Hanuacast itself can — or could — produce as many lead weights as needed in the country for incorporation in fishing nets. And Hanuacast is based in the same city as the country's largest manufacturer of nets. Other foundries could not rely on lead weights as a large source of income. Volumes of scrap — not just in accumulated dumps — but as a continuous flow would have to be judged. And if income were to be earned like Hanuacast through export sales to Australia, transport costs would have to be measured. It may be, for example, that aluminium ingots made in Mt. Hagen could not economically be taken by road to Lae and then by sea to Australia. In sum, if Hanuacast, with its favoured location is operating to such close financial margins, other foundries in less favourable places may not be viable. At least, they may not be viable if designed to the Hanuacast model.

There may in fact be better prospects for replicating the Hanuacast model in other countries. Visits by Non-Ferral representatives to New Caledonia, Tonga, Fiji and elsewhere in the Pacific region have indicated that supplies — and flows — of non-ferrous scrap do exist. From time to time they have received shipments from these places. But poor sorting of the scrap has meant that Non-Ferral have not been able to give top payments and poor packing has meant high transport costs in relation to the value of the contents. In these places, then, the necessary spark or combination of circumstances has not yet occurred. Again, however, no assumptions can be made about technical or financial viability of Hanuacast-type operations. Scrap supplies, fuel availability, labour availability and cost, shipping costs and other factors must all be carefully studied. But at least

Hanuacast provides a reference against which new possibilities might be judged.

Appendix I

A REPORT ON THE FINANCIAL SITUATION OF HANUACAST AS AT 17TH AUGUST '84
By SPATF

This report was carried out because there seemed to be some confusion about the exact financial position of Hanuacast: from a brief look at the accounts it looked as though Hanuacast should be fairly healthy and, indeed, Peter Thomas, an English VSO acting as manager, increased the levels of staff because of this belief. However, there has been some unease at SPATF because no profits seemed to be materializing even while making allowance for the cash flow difficulties which Hanuacast experiences.

I therefore had a closer look at the accounts and found that Hanuacast has in fact got itself into debt to the tune of K19,250. This is a combination of a bank overdraft of K5,129 and a debt of K14,127 to SPATF which represents the reimbursement of wages over the past few months. This extreme situation has been partly caused by payments for two July shipments (27 & 28) being delayed. Normally 60 per cent of the value of each shipment is received roughly one week after it arrives in Sydney but these two shipments were held up by strikes. The 60 per cent portion should arrive during August and is expected to be in the region of K9,000. Two further shipments have also been sent during August which should also help the situation. In fact, if the expected values of the shipments are correct, a large proportion of the SPATF debt should be cleared by the end of August.

These delays with payment were thought to be the main reason why Hanuacast was experiencing difficulties but in fact there is another, more important reason. This is that Hanuacast is failing to break even by a small amount each month. In fact, there are only two months, January and April, which show a surplus and on average, spending is K1,000 more than the average monthly income.

This situation has not been notified before for two reasons:
a)   SPATF has not been sending Hanuacast a monthly invoice for the reimbursement of wages. In turn, Hanuacast has failed to record a realistic estimate of the amount due. Although they were aware that they must be owing something, the size of their debt to SPATF came as a complete surprise.

b) The second reason for Hanuacast not realizing their true position was due to a small, but significant book-keeping error. When the scrap is weighed for shipment an estimate of the final value is made. This estimated value was then entered into the books as expected income (under debtors). The actual payment from Australia, however, is always less the value of the freight which was charged on the shipment. In the past this has been around K1,000 for each container. This cost has not been entered at all in the books presumably because it was not a direct payment. However, this has meant that the level of debtors (and anticipated income) has been inflated each month and Hanuacast has been working under the assumption of these higher figures.

Having looked at why the true situation has not been apparent before we must now try to look at why Hanuacast is producing a small deficit each month. Below I have suggested several factors which might have a bearing on this. Some of these factors can be acted upon but others, such as the reduced metal prices, are beyond Hanuacast's control.

1. Frequency of Shipments:

These vary from an interval of 1½ weeks between shipments to an interval of 5 weeks. This seems to be partly due to the availability of scrap but there also seems to be an element of slack production. Peter was recruited initially as the Manager of the Foundry but, due to other pressures, he has had to take on the running of the whole of Hanuatek over the past year. This has left Wilson in charge. Wilson has made a very good job of it and is very talented but he still lacks some of Peter's experience and authority (the latter because he is regarded to be 'on the young side' by some of the other foundry workers). In addition, he too has been very caught up with book-work and has not been able to spend much time in the foundry. Foundry workers are therefore often left unsupervised and it is significant that both Peter and Wilson agree that production rises when one of them is present. It seems likely, therefore, that once a Manager of Hanuatek is appointed and Peter returns to the foundry, that production will be raised. Hopefully this should bring the gaps between shipments down.

2. Scrap Prices:

The payments for scrap seem to be rather disproportionate to the

amount of income recouped from it and I would suggest that two
things ought to be looked at:

● whether any fiddling is going on at the weighing stage. This could
perhaps be checked by one person being made responsible for all
weighings while occasional spot checks are made on that person
(especially when wantoks are bringing scrap in).

● to see whether the amounts paid for scrap are realistic. These
were not in fact costed out at the beginning and in fact it appears
as though the rates might be too high, especially for some
material such as aluminium fly wire. I would think that it is
therefore a priority to look at costings and change prices if
necessary. Even if prices have to be lowered I don't think the
availability of scrap will be too badly hit since little has been done
by way of advertising or contacting businesses which might want
to get rid of their waste.

3. Metal Prices:

Since Hanuacast was started some of the prices for metal have fallen
instead of risen. Copper, for example has fallen from a rate of 1.20
dollars per kilo to 0.90 per kilo. This, combined with the effects of
inflation, has meant a slight drop in income.

4. Additional Costs:

In the past few months some of the costs incurred have been
unnecessarily high. For example, there have been times when the
recycled scrap has been too high an amount for one shipping
container. Therefore, in addition to one full container load (FCL) a
lesser container load (LCL) has been sent out. However, the same
handling costs and fumigation charges are still made on an LCL as
an FCL. They are therefore disproportionately high for the value of
the contents. Hanuacast has therefore recently changed its policy to
one of only sending FCLs which should reduce costs in future. They
have also succeeded in getting a reduction in their freightage from
K1,000 to K800. This should produce an annual saving of around
K3,000.

5. Costing of Activities:

Not much thought seems to have been given so far to the cost
effectiveness of activities at Hanuacast, i.e. comparing the costs of
various activities and the income they generate. For example, it

would seem that castings produce far less revenue than smelted scrap, yet the production of the castings depends upon the use of the furnace. If you think in terms of opportunity costs you can see that the real cost of the castings is in fact a lot higher than the paper one, since it is blocking the *opportunity* to generate higher income. It is perhaps arguable that some decision needs to be made about the aim of Hanuacast. If the aim was to break even or to make a profit then the less productive activities should perhaps be dropped. If however, the aim is to provide a training ground where people can learn some of these skills then perhaps some thought should have been given to subsidizing these activities.

Having said that, it looks as though the problem will probably sort itself out. Once Peter returns to Hanuacast he intends to build another furnace. This should allow both smelting and casting to be carried out simultaneously. However, this still does not get rid of the need for a decision on the exact role of Hanuacast. Should it just be covering its costs, be a revenue earner — or what?

Cash flow budget

For the sake of this projection I have assumed that Hanuacast is aiming to produce a small profit each month. This will need a few changes to be made but, after talking with Peter and Wilson, it seems that they are realistic. I have therefore prepared the budget assuming that some of the most important changes are made:
- That production is raised, and kept, at the level of one shipment every three weeks.
- That the average value of each shipment should be K7,900. This is roughly what the average gross value is at present.
- That costs remain reasonably stable over the next few months. This may mean that Hanuacast has to examine its expenditure very carefully and cut down on unnecessary expenses.
- That a Manager is recruited from Hanuatek and that Peter returns as Manager of the Foundry by the end of September.
- That another furnace is built in the foundry. It has been estimated that this will take about three months to complete. It is therefore unlikely that there will be any significant benefits from it during this budgeted period, but I have made some allowance for a slight increase in the income from castings towards the end of the year.

Can I emphasize here that all the figures in this cash flow are estimates. The picture could be much rosier if shipments are of a higher value than expected. The opposite is also true — it could take Hanuacast longer to get on its feet if production does not rise as well as planned.

It is now a question of, "Where do we go from here?" Hanuacast has not been making money but it certainly looks as though it has the potential to do so, especially if some of the changes mentioned in this report are carried out. I would think that the cash flow problems will remain for some time since Hanuacast will carry on being affected by delayed payments until it has built up sufficient funds in the bank to cover itself. Some support will be needed until this position has been reached.

At this moment this support is coming from a bank overdraft but this means that Hanuacast will have the extra costs of interest charges to bear. I would suggest, therefore, that now we know the exact position of Hanuacast SPATF should perhaps think about lending them enough money to clear their overdraft. This would mean that the total debt to SPATF at the end of August would be around K11,000 comprising of K5,129 from the overdraft together with the balance of reimbursed wages which has been estimated at K6,107. This debt would be reduced gradually over the next few months and I would see it being cleared at the beginning of next year.

It is also arguable that SPATF should perhaps continue to pay Peter's salary, even after he returns to Hanuacast. Peter is, in fact, an 'extra' at the foundry — once he leaves he will not be replaced. The salaries being paid at present, from Wilson down, are therefore the only legitimate ones on which the foundry should be judged. If SPATF decides upon this step the cash flow projection would improve slightly.

Whatever decisions and changes are made, Hanuacast needs to monitor its monthly progress to ensure that it is going along the right lines. This will be a matter of filling in the actual figures into the cash flow budget and to compare these with the estimated ones. This means that if, for any reason, the actual figures are lower, the situation can be picked up immediately and the appropriate action taken.

Dorothy McIntosh
Management Training Officer
August 1984

Appendix II

AN ALUMINIUM FOUNDRY IN
PAPUA NEW GUINEA

Reproduced from *Appropriate Technology* Vol.12 No.4
March 1986.

*Shortly after this report was prepared Hanuacast moved to a new site with
more land. Under the guidance of Willie Feinberg, an improved furnace was
built giving greater scope to the company's activities. The article below
reports on Hunuacast's position in the months following.*

Hanuacast is a small company in Papua New Guinea's capital of
Port Moresby, set up five years ago to take advantage of the
volumes of non-ferrous scrap accumulating as a by-product of the
country's industrialization. After four years during which the
process of smelting scrap aluminium and other metals was slowly
mastered, the company was in a position to move to a new site
where a greatly improved furnace had been built. With the addition
of a sand-pit for making aluminium castings, Hanuacast is starting
to live up to its original high goal of producing castings for local
industry.

Up to 400 people from squatter camps on the edge of the city
are earning an income from collecting and sorting scrap. They sift
their way through two of Port Moresby's large rubbish dumps and
reclaim scrap materials generated by new construction projects, old
motor vehicles and the general residues of industrial development.
One reason for the popularity of the Pacific Basin for scrap recovery
used to be the availability of abandoned war materials. Most of
these have long since been appropriated but Hanuacast did manage
to locate a World War II Japanese aeroplane which was one of the
first sources of metal for the new foundry.

Modest beginnings

The project was started in 1980-81 with the help of ITDG by
Hanuatek, the village technology arm of the South Pacific
Appropriate Technology Foundation (SPATF), when the huge
market for scrap metals in Australia was recognized. ITDG
technical adviser Willie Feinberg built a furnace of simple

building bricks with a refractory lining of clay, sand and cowdung for only US$250. Aluminium was melted down from alloys and cast into 20kg ingots, formed in moulds made from lengths of angle iron welded to form a trough. Other types of scrap, such as copper, brass, bronze and stainless steel, which could not be melted down, are sorted, packed into 200 litre oil drums and shipped direct.

As the foundry workers gained experience in distinguishing between different types and grades of material, and gave more attention to the selection and smelting of alloys, Hanuacast was able to get better prices. Early shipments were often of low-grade materials, and aluminium ingots contained high proportions of contaminants, particularly zinc. This affected the price the Australian buyer was prepared to offer, as further refining work was required. As the workers gained experience, Hanuacast was able to earn top rates for the quality of its shipped aluminium. On the buyer's own admission, the purity of aluminium is as high, if not higher than any being recycled in Australia.

Total investment in the new foundry amounted to $20,000 but, with shipments of up to $100,000 per annum of aluminium and other scrap metals, and the extra income from aluminium castings, Hanuacast expects to repay its loan in less than two years. Wilburt Feinberg visited Papua New Guinea again to design the new foundry and supervise the building of the furnace, which is made from a double wall of refractory bricks and furnace mortar, closely set to prevent gases escaping. Once it is heated to 660 centigrade, the melting point for aluminium, the molten material is funnelled through a channel in the bottom of the furnace into moulds which are laid onto steel rollers for easier, more efficient handling. At this temperature, other metals such as copper, brass or iron are not melted. These and other impurities are either scooped out of the molten aluminium or left in the furnace to be raked out after each batch of scrap has been processed.

The furnace is fired with waste motor oil, with a blower funnelling air through the oil nozzles. Waste sump and gear oil is available free of charge from garages and public agency vehicle service depots, which are forbidden to dump waste oil, although there is no public disposal service. This is a rare occurrence of legislation working by accident in favour of an appropriate technology venture. A characteristic of this fuel is its high content of carbon which effectively cleans the aluminium as it passes through the furnace, resulting in ingots of high-grade 95 to 98 per cent purity. The furnace is operated on average three days a week on a round-the-clock basis. Since it takes two hours to bring the furnace up to temperature, it is more economical to keep it running once it has been started. This mode of operation will also help to extend the life of the furnace. Wear and tear to the furnace is caused each time it is fired, as the bricks expand and contract with changes in temperature.

Sand castings

Now that its 10 foundry workers have become adept at handling the technology, Hanuacast is diversifying into aluminium castings for the local market, one of the original aims of the project. Already the company is filling orders for connectors for electric pylons and orthopaedic arm-crutches, at less than half the price of Australian imports.

The high-grade aluminium produced in the foundry is ideal for making castings but an early obstacle was a lack of sand with the correct properties. The material used in the first trials was sticking to the castings, giving them an uneven finish. Willie Feinberg resolved the problem by prescribing a mixture of locally available washed sea sand and river sand, high in silica. This is combined with waste engine oil, with the excess oil burnt off as gas and liquid, leaving a sticky substance with good binding characteristics. A sand pit has been included at the new foundry site for making the castings.

To make a casting, sand is packed tightly around a wooden or plaster model of the desired shape in two symmetrical, interlocking halves. When the mould is set by the patterns, the two halves are separated to remove the pattern, and then re-linked together. A small hole or sprue is made in the sand mould, and molten aluminium is poured into it. When the metal has cooled and set, the mould is opened and the sand broken away from the casting. The company is now working with lathes, shapers and millers for finishing its products. These previously unused tools became available from SPATF after the failure of a small-scale industries' project.

Hanuacast has other castings in the pipeline, for instance, aluminium/bronze castings for the heads of water pumps which are being installed as part of a WHO project in PNG. At present, the heads are made from galvanized mild steel, with a number of reducing adaptors. Apart from the cost of these items, the threads on the adaptors are a breeding ground for disease. Made from a cast alloy at the foundry, each unit can be formed in one piece, making it cheaper, more sanitary and easier to install and maintain.

Lost-wax casting

The company is also hoping to move into the field of lost-wax casting in the near future, an area in which Willie Feinberg has a great deal of experience.[1] The first area of application could be in making pelton wheels for micro-hydro projects being established in PNG. With the foundry's low operating costs, Hanuacast estimates that it could make pelton wheels to a high standard for as little as US$5, providing a substantial saving on imports from America.

The success of the Hanuacast project has attracted a great deal of interest from other islands in the Pacific for the idea of scrap reclamation. Feasibility studies will be carried out in the near future through the Foundation for the South Pacific to examine the potential for similar small-scale projects in Tonga, Fiji, the Solomon Islands and Honiara.

by Ian Macwhinnie, ITDG

Reference
1. W. Feinberg, *Lost-wax Casting: a practitioner's manual, IT Publications, 1983.*

Appendix III

DRAFT PROPOSAL

Introduction

It is proposed that SPATF, the Foundation for the South Pacific and ITDG cooperate in order to promote and develop small-scale scrap processing, recycling industries and small-scale non-ferrous foundaries in island nations of the South Pacific.

Background Information

1. Over the past few years SPATF has developed the technology to carry out the processing of a range of scrap metals using locally fabricated equipment. Capital costs are low, as are running costs since the furnace uses waste oil as a fuel.
2. An efficient streamlined scrap sorting, processing, and packing process has been developed, run effectively by nationals of relatively low educational level. Simple procedures for scrap purchase, equipment maintenance, accounting, marketing and overseas shipping have been effectively implemented, again run totally by Papua New Guinean staff.
3. The above mentioned process, run by Hanuacast, a division of SPATF, is profitable in its own right as well as having many benefits to the community including direct employment and income earning opportunities to urban settlement dwellers who collect and sort scrap.
4. The major buyer of sorted and processed scrap from Hanuacast has stated that quality is higher than many of their suppliers within Australia and that they would be interested in purchasing any amount from other countries of the Pacific.
5. There has been some development of further processing of the scrap into finished products of lead and aluminium for import replacement within the country or for export. There is much scope for development here, although the English volunteer assisting SPATF is soon to finish his contract with them.
6. ITDG has expressed an interest in supporting further research and development into the processing of scrap metals and other waste materials in developing countries. They see that such developments create income earning opportunities among otherwise disadvantaged urban dwellers.

7. ITDG would be in a position to disseminate proven technologies, as developed in PNG, elsewhere in the world. As has occurred elsewhere ITDG may also coordinate and support skilled personnel to assist in respective countries developments.
8. FSP has a presence in many island nations of the Pacific and through its small project grants is intimately involved in the development of income earning opportunities in agriculture, industry and commerce in small village and urban communities.
9. FSP would support any new projects that fulfilled the above stated objectives, especially if the projects have been proven viable elsewhere in the region and qualified personnel were available in nearby countries.

Proposal

a) That SPATF provide the facilities to allow further research, development and refining of technology in regard to scrap processing and the small-scale production of non-ferrous cast products.
b) That ITDG support such R&D through the funding of or contributing towards the funding of a person to carry out this work.
c) That FSP undertake to support and assist in the establishment of units similar to Hanuacast in other centres in PNG and in other countries of the Pacific.

Advantages

(i) To ITDG: increased knowledge of a particularly useful level and type of technology which has application in many other countries of the world, and an increased presence in Pacific countries where previously they have had little input of any sort.
(ii) To FSP: promotion of a proven, viable, profitable small-scale industry that has low capital and running costs. Applicable to many Pacific countries who may have few natural resources.
(iii) To SPATF: improvement in existing scrap recycling, processing and casting facilities. Developments of new possible import replacement products, and an increased representation in some form in other Pacific countries.

John Brooksbank, Dept of Industrial Development

69

www.ingramcontent.com/pod-product-compliance
Lightning Source LLC
Chambersburg PA
CBHW072155020426
42334CB00018B/2014